Enduring Acequias

Querencias SERIES

Miguel A. Gandert
and Enrique R. Lamadrid,
Series Editors

Querencia is a popular term in the Spanish-speaking world that is used to express a deeply rooted love of place and people. This series promotes a transnational, humanistic, and creative vision of the U.S.-Mexico borderlands based on all aspects of expressive culture, both material and intangible.

Also available in the Querencias Series:

Chasing Dichos through Chimayó by Don J. Usner

Hotel Mariachi: Urban Space and Cultural Heritage in Los Angeles by Enrique R. Lamadrid and Catherine L. Kurland; Photographs by Miguel A. Gandert

Sagrado: A Photopoetics Across the Chicano Homeland by Spencer R. Herrera and Levi Romero; Photographs by Robert Kaiser

Enduring Acequias

WISDOM OF THE LAND, KNOWLEDGE OF THE WATER

≈

Juan Estevan Arellano

UNIVERSITY OF NEW MEXICO PRESS • ALBUQUERQUE

Library of Congress Cataloging-in-Publication Data
Arellano, Juan Estevan.
Enduring acequias : wisdom of the land, knowledge of the water /
Juan Estevan Arellano. — First edition.
        pages cm. — (Querencias series)
Includes index.
ISBN 978-0-8263-5507-2 (pbk. : alk. paper) — ISBN 978-0-8263-5508-9 (electronic)
1. Irrigation canals and flumes. I. Title. II. Series: Querencias series.
TC930.A74 2014
333.9100972—dc23
2014001713

COVER PHOTOGRAPH: Rich Reid, courtesy of Getty Images
COVER AND INTERIOR DESIGN: Catherine Leonardo
Composed in Minion Pro 11/14.5
Display is Meridien LT Std Medium and Frutiger LT Std

Dedicated to the memory of my mentors, who knew what
it meant to have such a dignified title as mayordomo de la acequia,
none of whom are with us to guide us: Cleofes Vigil from
San Cristobal, Andres Martinez from Cañon in Taos, and
Pablo Romero from the Acequia del Llano in Dixon.

Contents

≈

Acequia by the road. Photograph by the author.

Acknowledgments

≈

This work had its beginnings, unbeknown to me, when I was a child playing in the acequias in Cañoncito and seeing my parents grow so much food, from chile and corn to an array of sweet cherries, apricots, and other fruits. Then for a while they blurred from my vision during my time in college; when I came back from my fellowship at the Washington Journalism Center and started to work with the Living Lab program for HELP (Home Education Livelihood Program) in Peñasco, acequias were once again front and center. It was there that I was introduced to oral history, working with Facundo Valdez and Dr. Tomás Atencio, who at that time were instrumental in beginning La Academia de la Nueva Raza. Though I was working under Facundo, a lot of the work I did was in helping to start the Academia archives with Tomás. Little did I know this was going to be my life's work. Along the way I met many people throughout the villages of northern New Mexico, and nothing was more important than land and water, the land grants and acequias. Along the way I met the Reverend Antonio Medina, and today I still continue working with him through the New Mexico Acequia Association, especially since our trip in 1998 under the auspices of the Ford Foundation with Dr. Walter Coward, who helped us with a grant to take nineteen New Mexicans to visit Spain for three weeks. There we saw the acequias from a different perspective and experienced a different type of artisan agriculture, one I was unfamiliar with. All of this added fuel, bringing me to eventually one day write a book about acequias and similar community water systems.

In Spain I realized that that country had a long history with acequias, all part of our history in New Mexico. Then, during my tenure with the Oñate Center in Alcalde, one of the first workshops dealt with acequias. While there we had many events dealing with acequias and land grants; 1998 was a watershed year since it was the Cuarto Centenario of Spanish settlement in New Mexico and the 150th anniversary of the signing of the Treaty of Guadalupe Hidalgo (thanks, Celina Garcia, Joyce Guerin, and Liddie Martinez). These historic events opened the door, through the Camino Real de Tierra Adentro, to visiting most of Mexico. There I met people who later became important in doing this work, among them Dr. José Rivera, and I again reconnected with Dr. Enrique Lamadrid when we worked on a children's book on acequias for the University of New Mexico Press. Rivera then introduced me to Dr. Tomás Martinez Saldaña from the Colegio Postgraduado in Texcoco, Mexico, and also Dr. Thomas Glick from Boston University and Dr. Luis Pablo Martinez from Murcia, all of whom have been very instrumental in completing this work. Dr. Gary Nabhan has also been a great resource and friend. Thanks also to Mylene d'Auriol for her wonderful photographs of Peru. Also Donatella Davanzo and Ellen Fowler.

I am not a historian, though I am a student of history, so this work is more in keeping with my vocation as a journalist, trying to tell a story of the common men and women who work the land, unnoticed and unrecognized—the faceless humans who have created civilizations, never asking for anything in return. Among them have been the hundreds of acequia mayordomos, commissioners, and *parciantes* who have toiled in obscurity, maintaining these grand systems that provide water to feed their villages, whether in New Mexico, Mexico, Spain, the Middle East, Asia, or anywhere in the arid world where people depend on these systems to survive.

Of course this work could never have been completed without the help of my wife, Elena, who has always been there for me, and my sons, Javier and Carlos, and daughter, Única. I also want to thank the Christensen Fund for their financial support through the Lore of the

Land (thank you Suzanne Jamison, a true friend since the days at NMSU, and Jack Loeffler) and the New Mexico Acequia Association (gracias Paula Garcia, Janice Varela, and the rest of the staff). And of course there are many others; you know who you are, but due to space limitations, I can't list all the names.

Canova de Trampas. Photograph by the author.

PART 1

THE

WISDOM

OF THE

LAND

La tierra dirije al agua, y el agua guia la tierra.

The way the Indo-hispano looks at the land can be found
in documents from cultures from the Middle East, Spain,
the Mediterranean, and Mesoamerica. Among those documents are
the Siete Partidas, the Ordenanzas of 1573, the Laws
of the Indies, and the Plan of Pitic.

Introduction

≈

MIS MEMORIAS, RECUERDOS

El olor a humedad y frescura,
el agua repleta de vida . . .
el reflejo y las sombras . . .
todo alrededor cambiaba . . .
y pensar que a unos pasos el desierto y la arena . . .
ardiente y seco contrastaba con la apacible acequia
que esperaba.

—ANONIMA

The humid smell and freshness,
the water full of life . . .
reflection and shadows . . .
everything changes . . .
and to think a few steps away desert and sand . . .
burning and dry contrasting with the peaceful acequia
that waited.

—ANONYMOUS

MIS PRIMEROS NUEVE MESES LA pase en la barriga de mi madre, sin pena ninguna siendo que la pasé nadando en el agua, y cuando vine al mundo nací como 200 yardas del río Embudo en la resolana de los barrancos blancos, que saltan del remance de la sirena cerca la Bolsa, unos pasos del río Grande en el lugar de la oscurana cerca la Junta de los ríos. Todavía

Desagüe, Embudo. Photograph by the author.

vivo en el mismo lugar que durante la tardeada los dos cerros hacía el
poniente parecen dos cenos de una joven, viendose por entre las corti-
nas de las ojos del arbol de albaricoque, en una postura de yoga, donde
solos los cenos saltan.

THE CHANGING LANDSCAPE
OF COMMUNITY AND WATER

This work, whatever it might be, is a writing experiment, incorporating (1) research—archival, oral history, genealogy, and personal history; (2) travel experiences—throughout the Río Arriba bioregion; Mexico, from Juarez to Chiapas; and Spain, from the Basque Country to Andalusia; and (3) practical experience, since I was born into a family that always lived off the land and we have continued that tradition with the creation of our own experimental space in a harsh high-desert environment, a combination experimental farm and recreational site that I call my *almunyah*, from the classic Arabic word meaning desire. This space is what anchors me to the land, a land where water is the most valued resource. Few people have learned to use water as wisely as those who rely on the acequias, open-air water canals common throughout the arid world; these people have developed a philosophy of sharing water that applies globally. To understand one acequia, the acequia that provides our food, our water, life itself, we must go back in history and around the globe in order to understand how it works.

"In this book," Roberto J. Gonzalez wrote in *Zapotec Science: Farming and Food in the Northern Sierra of Oaxaca*, "I attempt to combine both local and global approaches by maintaining a strong focus on one village while exploring global events through history and prehistory. In other words, I have tried to view global changes from a local vantage point." I will do the same. Whereas he focused on Talea in the region of Rincón in the Sierra Juarez in the Mexican state of Oaxaca, my focus will be on the Acequia Junta y Ciénaga within the land grant of Embudo in New Mexico, part of the Embudo watershed, where the Sierra Jicarita reigns as queen. This will be for the understanding of place, or in Spanish, *querencia*, love of place.

Acequias are what give us a sense of place, and the water becomes the blood that brings communities together, that separates the commons from the *suertes*, a land division introduced by the Spanish Crown, while at the same time uniting and making the land grant landscape one. But the acequias were not built for modern-day watercolorists to paint; they were a necessity for survival, for without the acequia water there would be no food.

My journey to reconfigure my space began in 1987 when my only daughter, Única Paloma Lucía, was born on a beautiful April 10 afternoon. My attempt here, as in my hectare, is to create a collage with words and photos, based on my memory and the memory of the land that has molded me, showing what it is to live in a rural space in northern New Mexico inhabited by ghosts and memories of my ancestors for centuries. Every time I step outside I'm in communion with those who worked the land before me. I am reminded of their presence by all the pottery shards found throughout and by the whistling of wind as if they are playing a flute: *al pasar el cementerio me chifló tu calavera* (as I passed by the cemetery your skeleton whistled). For the landscape carries the memory of those who came before me, those who also understood that you cannot separate water from community.

It's my family's odyssey, but more than that it's a person's journey in search of querencia, of breathing and living querencia, of defining querencia, both with words and with pick and shovel, with poetry and by planting trees. *Querencia*, "place": love of place, that sense of place defined by the texture of biting into a recently plucked green chile, the smell of tortillas cooking over a piñon fire on my grandmother's old wooden stove, the color of a ripe tomato waiting to be sliced. *Terroir*, the French call it; to us it's querencia. From the town of Arellano in the Basque region of Navarra, to Aguas Calientes in central Mexico, to Embudo, my ancestors crossed the ocean and traversed the hot desert terrain of the Jornada del Muerto along the Camino Real de Tierra Adentro—the road of water, one might say—for me to arrive at La Junta de los Ríos, the juncture of the Río de Picurís, now the Río Embudo, where I have come to anchor my bones, where I have found my querencia. As humans we are at a juncture—*la junta*—where we can survive sustainably or destroy Mother Earth due to humanity's greed, all in the name of God.

But *la junta* also has a different meaning, the gathering or coming together, the same as an *embudo* is also a funnel, where everything is gathered before it embarks on its journey; in a sense embudo and junta are one and the same, for before water goes through a funnel it has to come together. It is here that I have come to learn the secrets of the ancient acequias, but especially one, the Acequia Junta y Ciénaga, or the "ditch of the juncture and marshland," which quenches the thirst of my land, my plants, my trees, my animals, and my family.

Long, long ago, in the kitchen of the late Demostenes Griego—a descendent of the early Greeks who came to northern New Mexico and because of their unpronounceable last names were referred to generically as Griegos, the Spanish word for Greeks—when I was about nine years old in the midfifties of the last century, I remember my dad and his *vecinos* (settlers/neighbors) smoking home-rolled Prince Albert cigarettes while they discussed acequia business. Then, during the day, I remember going swimming in Cañoncito in the Acequia del Medio that ran through the center of Arellano land (which I later learned was actually Martinez land) with my cousin Anita, who was three years older than me. By this time we had moved to La Junta, about five miles west. It seems like I was born in the acequia, because ever since I can remember, the acequia has been part of my life, and the older I've gotten the more entrenched it's gotten into my blood. It is a memory of a certain landscape that invades my dreams, tortures me when I am awake, knowing that in a generation or two this landscape will be a thing of the past. Today I continue this journey in search of the story of the acequias, trying to understand water in the context of an old system based on community and to retell that story to a new generation. For a long time, like a lot of *nuevomexicanos*, I thought we were the only ones who had acequias. But I have come to find out that the type of irrigation provided by acequias might have originated in the Bronze Age civilization of the Indus Valley. But at almost the same time, though maybe a little later, the indigenous people of what is now Peru and the southwestern United States and other desert people were irrigating their crops using the same type of system, though under a different name. The Pueblo Indians of Ohkay Owingeh (Village of the Strong People)—which was renamed San Juan de los Caballeros under the Spanish Crown in 1598 and only recently returned to its original name—called this type of irrigation by canal *kwi onu, kwi on.*

But I am also interested in finding out how these acequias are tied to the land, the land our ancestors called *mercedes*, or land grants. A merced is a gift, but the gifts I am after are the gems of wisdom imparted by learning from the land—what we called in La Academia de la Nueva Raza *el oro del pueblo*—and the knowledge revealed by the water as it meanders from one bank of the acequia to the other, creating its own journey, in search of that sacred knowledge we all aspire to find.

I remember as a young kid going from Cañoncito, where we lived until I was seven years old, to Ojo Sarco, where my dad's youngest sister, Merced, used to live, and how on our way back through the Cañada del Oso my dad would say, "Todo ésto le pertenece a la mercé" (All this belongs to the merced, to the grant). I thought he meant it belonged to my aunt, and I would say to myself, "My tía Merced sure has a lot of land."

On our way to Ojo and on our way back, or when we would go for wood, we would always stop for a drink from Ojo del Oso, Bear Spring. There would be tracks of deer, coyotes, and an occasional bear. The spring had the most crystalline and pure water I have ever tasted. One could count the grains of sand on the bottom of the *chupadero*. A chupadero is a small bowl made by hand to gather water so one can drink from it. When water gathers in basalt eroded by nature, the rock formation is known as a *tinaja*, or bowl. Now this type of traditional knowledge is a forgotten nugget of gold. People can be stepping on top of an unclean chupadero or next to a tinaja full of rainwater and die of thirst. This water is a lot better than the $1.80 bottled water the local store imports from Tuscany. It's a cooperative that espouses sustainability, but still the water is imported from Italy! If people only knew about this springwater and rainwater, they wouldn't waste their money.

As an adult I found out that this same *cañada*, or arroyo, was the eastern boundary of the Embudo land grant, granted to my ancestor Francisco Martín and two others in 1725. To me, then, a merced didn't relate to my querencia, or sense of place, but to my wonderful aunt, who

La Jicarita. Photograph by the author.

would always hug me and kiss me whenever she saw me. That was *familia*. And when she came to visit she would bring *calabazas mexicanas*, *chicos*, posole, whatever she had, as part of the *convite*, the sharing of food tradition that disappeared with the coming of supermarkets.

Later, as I grew older, during the summers Aaron Griego, our Little League baseball coach, would take us up to the lakes under Truchas Peak, which form the easternmost part of the Embudo watershed; though at that time I wasn't interested in learning about the watershed or water. We never took water on our seven-mile hike up to the lakes, for there was plenty of water running on the creek, populated by *panzas coloradas*, red-bellied native trout. The Rito de San Leandro could be heard gurgling a few feet beneath the trail going up. If we got thirsty we would make a cup with our hands to drink the cold, pure water from Laguna Escondida—today Hidden Lake—under Truchas Peak, for here water oozed from a spring under a rock sculpted with all the details and finesse of a Henry Moore sculpture, where the wildlife also gathered to quench their thirst. We always knew where to find water.

My father would always tell me, if there's sand, there's water; the same if you see cottonwoods—there's water, because they have shallow roots and they grow only where there's water close to the surface. In the fall one can make a mental map of the places with water simply by noticing where the cottonwoods are growing as they turn from green to a golden yellow.

Therefore, this work in search of the acequia trails begins many centuries ago (hundreds of moons into the past), and not only in the Americas, for I am Apache and Pueblo, but also on the Iberian Peninsula, high in the Pyrenees Mountains; in the deserts of North Africa near the Holy Land and of the Arabian Peninsula in southern Yemen; and even farther east, in the Indus Valley, in present-day Afghanistan and Pakistan. I am told by those who know that I might also be Sephardic/Basque and Moorish. What a combination: an Apache-Basque and Arab-Jew, settled in Anasazi land, who speaks a language we call Spanish, although it is not Castilian (made up of one-third Arabic words), and more than a smattering of Nahuatl and a sprinkling of Hebrew and Latin.

Clarification: I am not Spanish; I consider myself New Mexican, for my father always said we were "mexicanos"; and I am not a Jew, nor am I a Moor, but I do think of myself as Chicano. Neither am I Basque, or Apache, or Picurís, but I am all of the above; I am Indo-hispano, whatever that might be. I was a global creature before *globalization* became a buzzword; I am a Heinz 57, a mestizo with my taste buds on several continents. I am a Chicano writer and a nuevomexicano, as the writer from Tierra Amarilla, Sabine Ulibarri (another Basque last name, I am told), would say. I have no trouble eating *buñuelos*, a Middle Eastern specialty, with chile and *maíz* from Mesoamerica and for dessert having *arroz con leche* from Morocco or *capirotada*, a Sephardic Lenten dessert. All while sipping a glass of Rioja wine with my dinner and an after-dinner drink of sotol or mescal, communicating in Spanglish with my family while I attempt to decipher a document on the *feixes* of Ibiza written in Catalán. I am a walking contradiction. This is me, this is my family; join me for a cruise around the block, or better yet, the *manzana*, which also means "apple" but here refers to a city block, or more precisely to 1.43 hectars. This cruise will take us across the globe and back to see how community and water survive as a divine right.

Water is not a commodity; it belongs to all living beings—humans, animals, and plants. I tend to follow the Law of Thirst, as it is known in Islam. My mother would say, "Para vos, para nos, y para los animalitos de Dios" (For them, for us, and for God's little animals). One can't find a better definition of the Law of Thirst. Water should not be sold for a profit.

My journey (that is, searching my roots) begins in the town of Arellano, Spain, near Estella, about eighty miles from Pamplona, the city made famous for the running of the bulls during the fiesta of San Fermín every July. The town itself is situated on a bluff, looking down on beautiful green fields known for their white asparagus, not too far from the Río de Oja, better known to wine connoisseurs as the Rioja. According to history, the town of Arellano came about in the mid-1300s, when a Sánchez Ramirez (he was given the land of Arellano and thus became the first Arellano) married the granddaughter of El Cid. Yes, El Cid, the killer of Moors! I am not claiming to be a direct descendent of El Cid, nor of Geronimo, though my father would always remind us that my great-great-grandmother María Albinita Maes de Martín (Martínez) was a full-blooded Apache who lived to be 105 years old. He said he met her only once, when my grandfather José Agustin Arellano sent him up to Trampas on horseback when he was about ten years old, on some errand that he didn't remember. All he remembered was that she told him, "Ven pa'ca mi'jito, arrimate a ver a quien pareces" (Come here, my son, get close so I can touch you, to see who you favor), because she was already blind, and that she had cried as she touched his face, like a sculptor molding a face. She was the mother of my great-grandmother María de la Luz, whom Silas Salazar from Cuestecitas called "la mamá de los Arellanos," the "mother of the Arellanos." María de la Luz married José Ignacio Arellano from Bosque Grande, today known as Canova, across the Río Grande from present-day Velarde. Before the post office was established it was called La Joya (or Jolla), meaning "the jewel" but also "hallow," because of its fertile lands. Previously the Arellanos had lived by La Villita near present-day Alcalde (La Soledad del Río Arriba); their property was by *la capilla*, "the little chapel." Albinita was married to Juan Isidro Martín(ez). She was born around 1832, according to genealogy research. If my father was correct, since he was born in 1903 and it must have been 1913 when he went to Trampas, she must have been in her eighties at that time.

On the Martín side there is also Pueblo blood, I was told by Santa Fe genealogist José Esquibel when I asked him if the Martíns had Tlaxcaltecan blood in any way. That's another part of the puzzle, which will be discussed later. He said he had found Tanoan blood in the Martíns. And "Gary Nabhan, himself with roots in Lebanon, has written that the Martíns" have Moorish blood: "Some of those refugees [from the Spanish Inquisition] had to conceal their surnames and their religious orientations from those who offered transport to the lands now known as Latin America. Some of them took code names for their *apellidos* [last names]— animal names such as León, Garza, Gallo, Martín, Ossa, Tigre, and Zorrillo—so that others of their kind could recognize them. Others simply shifted a few consonants in their surnames to obscure their origins."

He goes on to say that many of these families, "Catholic on the outside, Jewish or Moslem on the inside, . . . fled as far from the Valley of Mexico as they could when the Inquisition finally followed them to their newfound land. They took refuge in the most remote regions of New Spain—present-day New Mexico, Nuevo León, Chihuahua, Sonora, and Coahuila—as well as in the more remote haunts of the Yucatán Peninsula. In that first wave of immigration from Iberia to Mexico, roughly one-third of all immigrants were originally from Andalusia, where Moslems and Jews had formerly lived in the greatest numbers."

Henry Street of Ponderosa Winery told me during a visit to his place in 1998 that what today is the town of Jemez (for the Xemes tribe has been there for centuries) had been settled by a Martín(ez) who was an Arab. And looking at it from a historical perspective, who else would have risked their lives to cross the ocean if not the Sephardic Jews and the Moors, both groups having been expelled from Spain in 1492? Why would the Christians risk their lives, when they were to inherit the Moorish lands, though they knew very little about how to manage an acequia or what an *alquería*, the predecessor of a land grant, was? Little did they know of the hunger awaiting them for not knowing how to work the land that they inherited. Still, in 1998, in talking to an eighty-five-year-old stonemason near Chinchón, the town near Toledo in central Spain that smells of anise at all times, he said, "When the men left from here, never to return, the mothers cried to the day they died because they never knew if their sons had survived, if they had married, if they had children, grandkids that they never knew." I had never seen the reverse side of the coin, but what agony and suffering it must have been for these parents, especially the mothers. Then in the same breath he uttered, "And there were also a lot of Indians who were brought here and never returned to the other side. There are still mestizos here." Their mothers on this side of the ocean must have also gone to their grave wondering what happened to their sons.

For those of us New Mexicans who are Indo-hispano or mestizo, place, querencia, starts with don Juan Narrihonda Salazar de Oñate's contract with Viceroy Luis Velasco, signed on September 21, 1595, in which Oñate agreed to follow the "Ordenanzas de descubrimientos, nueva población y pacificación de las Indias," promulgated by King Felipe (Philip) II of Spain in 1573. From these ordinances came the Laws of the Indies of 1681; they have been called "probably the most effective planning document in the history of man-kind" and "the most influential body of urban law in human history" by Axel Mundigo and Cora Crouch in *Spanish City Planning in North America*. Also, keep in mind that because Oñate came from the Basque town of Oñati, which I had the privilege of visiting in 1997 and then in 1998 during New Mexico's Cuarto Centenario, his signed contract was the first attempt at economic development, and in retrospect a very successful one, for this group of settlers. Led by the Tlaxcaltecas, these settlers helped lay out the vast acequia system still in use in the Río Arriba (Upper River) bioregion and also established the livestock industry, the fruit industry, and the chile industry—because with Oñate came the first chile seeds from Mesoamerica (though Baltazar Obregon in 1580 claimed to have brought chile seeds with him). I know Native Americans don't like to hear about land grants, which they claim the Spanish stole from their ancestors, and I don't blame them. It is true, because before the coming of the Spanish Crown everything belonged to them. I deeply sympathize with them, but they must also realize that after four hundred years they can't continue calling us "Spanish," because we are not that either; we are a mixture of many bloods—we are mestizos! And whether they like to admit it or not, we are brothers and sisters, for there are very few on either side of the aisle that don't have either European or Native American blood. But sometimes scholars who are neither mestizo nor Native American continue attempting to drive a wedge between us so that they can have something to write about from their sterile ivory, or now adobe, towers.

Of course Oñate had many faults—he was human—and that's why he was eventually banned from New Mexico; he was a man *de hueso, carne y sangre*, "of bone, flesh, and blood," and he committed many atrocities against Native Americans, contrary to the ordinances. For that reason there is still a certain animosity among the Pueblo people against the "Spanish," though there were very few Castilians in Oñate's initial group of 129 families, made up of criollos, mestizos, and *gachupines*. Oñate himself was a mestizo, so he was not a *peninsular* (born in Iberia), and when he was banished to Spain that was the worst punishment that he could have gotten. Same as sending children born of Mexican nationals back to Mexico when the United States is all they have known since birth; to them this country is their querencia. And of course there are many who even today, four centuries later, swear they are "pure Spaniards," without realizing that Spain has always been a crossroads between Europe and Africa and later with America. Don't tell a Basque or a Catalán that he is a Spaniard; I had the misfortune of commenting to the alcalde of Oñati at a dinner in 1998 that the Spanish flag and New Mexico's flag shared the same colors. He then whispered in my ear, "We are not Spanish, we are Basque." Then I understood why my dad's friend Filigoño Sanchez used to say, "I don't celebrate the Fourth of July, I am not an American," and then would burst into a really loud laugh.

This work, then, is about horticulture, arboriculture, rural development, philosophy, and history; it's also a work about environmental art, spirituality, and landscape architecture, as well as poetry. But the underlying theme is water, *agua*, that which sustains all life, and how desert people have found that secret knowledge that allows life to thrive, if only with a few drops of water: that secret is sharing. The saying *agua que no ha de beber dejala correr*, "water that you are not going to use, let it flow," explains this philosophy very succinctly.

I recently came upon a thirty-minute video, *Landscape and Memory: Martinican Land-People-History*, produced by two professors from Bucknell University, Renée Gosson and Eric Faden, in 2003. In the accompanying brochure, they write, "In *Landscape and Memory*, the French West Indies' most renowned identity theoreticians—Jean Bernabé, Patrick Chamoiseau, and Raphaël Confiant—investigate the different ways in which France, as a colonial power, marks colonized lands and peoples. Importantly, this is one of the few films about Martinique that adopts a Martinican perspective on France's overwhelming and continued colonial and cultural presence." My son and I have been working on a film about acequias also, but the challenge has been to go beyond the tedium of most documentaries and try to examine the role of colonial power, so that Native Americans stop criticizing those of us who are mestizos as being insensitive and Spanish, while at the same time not denying who we are.

These filmmakers faced the same dilemma that I do, as they write, "The Martinican writers ask how, in a country (or a 'Department') like Martinique, does a colonial power 're-map' space and land? How does it 're-map' a people's memories and identities? And can one resist this re-mapping?" The writers, the brochure continues, "examine the possibilities of landscape as a repository for a forgotten past, Martinique's economic dependence on France, the recent 'cementification' of Martinique, the politics of commemoration, and the possibilities for Creole culture. The film combines the writers' environmental and ideological concerns with actual footage of the island, showing the symptoms of cultural devastation (satellite dishes, advertisements, supermarkets, regression of the mangrove swamp, etc.)." Substitute the place-name "northern New

Mexico," or rather, "Río Arriba bioregion," for "Martinique" and "mestizo" for "Creole," and the description seems like a seamless fit for what's happening in the place I call home, my "querencia."

This film was shot on location in Martinique in March 2001 by two independent American filmmakers. But most important for me as a writer is that

> *Landscape and Memory* also poses several questions about the documentary form. The film is called a *"média-stylo,"* paying homage to French film theorist Alexandre Astruc's 1948 manifesto *"La caméra-stylo."* This manifesto urged filmmakers to develop a genre that was neither documentary nor fiction but closer to the form of the essay—poetic, fragmented, open-ended, speculative, reflexive, and subjective. Using moving images, text, sound, music, and voice, *Landscape and Memory* is—to use Jean-Luc Godard's words—"research in the form of spectacle." By using this style, the film neatly reflects the structure of recent French West Indian novels, which are often themselves a métissage of history, narrative, documentary, and poetics.

For me this work is also a blending of history (oral, archival, genealogical), narrative (including folklore, which I refer to as traditional knowledge), and poetics. Hopefully it will also serve as the screenplay for a film on water, traditional agriculture, food, and community.

It's as I wrote in a poem from the late seventies, after visiting with an old man by the name of Miguel Armijo in Santa Fe's Alto Street barrio (an Arabic word meaning "community," appropriated by the Chicano activists in the 1960s), a man whose memory went back to the early 1800s because he had been brought up by his grandparents, who were very old when he was a young kid. After returning from Santa Fe I wrote the following:

A la niñez le gusta oir la historia,
    A la viejez contarla,
        Y a la juventud hacerla.

Children like to listen to history,
    The elderly enjoy talking about it,
        ·   And the youth like to create it.

It brought back memories of my youth, sitting at the side of my tía Josefa, a daughter from my grandfather Tomás Archuleta and his first wife, Marcelina, who by the time I remember her in the late fifties was quite old and blind. She was always dressed in long black dresses, with several *enaguas* (underskirts) where she kept her *punche* (tobacco), rolling paper, and matches, for she liked to smoke while she kept us entertained with her *cuentos folclóricos*, stories of old. She told us about Pedro de Urdemalas, who is known as the quintessential *pícaro*—rogue—throughout the Americas from Chile to New Mexico, a character that dates to the 1300s. Urdemalas is also known in Turkey, and Cervantes wrote about him in the 1500s. She also told us about witches, La Llorona, and other characters, for we didn't have TV or iPods to keep us entertained like today. Before she lost her sight, she and her husband, whom I never met, used to plant acres of chile and wheat, as well as her kitchen garden of vegetables, for the acequia went past her front door. She lived between the Río Embudo and Las Llomas de la Angustura. Now that place, like a lot of the land in Embudo, is abandoned and full of Siberian elms and other exotics. When my aunt had the place it was immaculate, it was a garden of paradise, for they knew the secret of water. After she died my cousin sold it to a doctor from Albuquerque, and ever since, for over thirty years, it has laid abandoned with only a few heritage apple trees surviving by the grace of God, for no one waters them and they still produce fruit.

Every year I stop to savor the tasty tart apples and say, "I have to graft these onto new rootstock and create new trees to save the varieties." I haven't done so, but that's on my to-do list. With the help of Ron Walser from New Mexico State University's Sustainable Agricultural Center in Alcalde and Gordon Tooley from Truchas in 2002, we started a project to save the old varieties of apples that came from the south up the Camino Real de Tierra Adentro. But due to lack of funds, we haven't continued with the project.

This work is part biographical; that is, the biography of a land grant, the Embudo land grant, and an acequia, the Acequia Junta y Ciénaga. It is this rugged desert landscape and the bountiful acequia that supplies the sweet water to our fields that have shaped me into what I am. And it's also autobiographical, since it's about my life, about my place—mi querencia—my relatives. I'll share one story with you immediately. My grandfather Tomás was born not Archuleta, but actually Borrego. My great-grandmother Ramona Archuleta, born in 1822, who I had been told never married, according to family lore, actually did marry twice. Recently, in some old family documents found at Princeton University, I discovered that when she was twenty years old she married Francisco Martín, a descendent of the original Martín clan, but I can't find the connection to the original Francisco "El Ciego" Martín, who was forty years her senior. They had three kids, but only one daughter, Alcaria, survived. After Ramona became a widow at age thirty in 1853, she married Juan Antonio Borrego in 1855; from here Tomás Aquino Borrego, my grandfather Tomás, who later became Archuleta, was born on March 7, 1856. Genealogist Henrietta Martinez Christmas provided this information; she is from Corrales. There was also another daughter, Benigna. One of my uncles used to say that my grandfather's father was a Borrego, so the family should have been Borrego instead of Archuleta, another Basque name. Borrego appears to have been Sephardic. But he also said that we were descendents of Francisco Martín. In other words, he was not sure whether he was Borrego or Martín. Archival research by Dr. Danna Rojo Levin, a professor at the Universidad Metropolitana de Mexico, recently uncovered a document in which Alcaria is listed as the daughter of Francisco Martín. Alcaria was born around 1853 (after her father died), my grandfather Tomás in 1856, and Benigna in 1858; Tomás and Benigna had the same father. The two half-sisters were married to brothers, Benigna to Donaciano and Alcaria to Matias Romero. Both Romeros and my grandfather were listed as landowners when the Acequia Junta y Ciénaga was registered with the Office of the State Engineer in 1915. Census records that Lorraine Aguilar, a teacher and genealogist, has combed through reveal that during the Navajos' Long

Walk in the 1860s, many Navajos—kids at that—remained in the Indo-hispano villages, including Embudo. And one of those households with a Navajo was my great-grandmother Ramona's. Also, my wife's great-grandmother Rosina is listed as a twelve-year-old Navajo in the 1880 census. Another layer of history never revealed, hidden from us, for the scholars tell us we are "Spanish," and we believe it. "You are who we tell you," they tell us, "not who you really are."

But more than genealogy, this work is about land and water in one very specific place, for to understand place, or querencia, one has to know the ground, the rocks, the trees, the flora and fauna. By knowing my roots, I have gotten to know the place *como mi manos,* "like my hands," as the old ones would say, based on the information stored in my mind and experienced through the senses, that repository of personal and collective memory. It's a very personal journey into the past, in search of answers for the future. In his Act of Taking Possession of April 30, 1598, the day of the first Thanksgiving in what today is this country, Oñate defined *querencia* as "from the leaves of the trees in the forests to the stones and sands of the river, and from the stones and sands of the river to the leaves in the forest." This hybrid work, then, deals with the history of the Embudo Land Grant and the *Recopilación de las leyes de los reinos de Indias* (the Laws of the Indies) and how this barren and unproductive landscape, through the means of artificial irrigation, was turned into one of the most productive pieces of agricultural land through the blood, sweat, and tears of my ancestors.

Now it's sad to see the land and communal water system go to waste in the hands of those who bought land because they could afford to but who have no ties to the land, no memory of its use; their umbilical cord was left somewhere else. They are aliens in this land, for they have left no sweat or blood working the land; for them it's only an investment. Now one even finds photos on the Internet of sacred places, taken by newcomers with digital cameras, and because a piece of land is vacant they immediately assume it's on the market and tag their photo "land for sale," advertising for more disconnected people to settle as if they were pioneers. That dreadful thing called money can never buy a sense of place; though they might write about a place or attempt to paint it, it's still about money. At the annual fiestas in Dixon—before the Plaza del Embudo—a band of die-hard white radicals march carrying signs calling for "Paz," peace in Iran, Iraq, everywhere except here. For some of these marchers, when things don't go their way with the water in the acequia, they immediately start complaining about the locals not knowing how to manage the water, when it is they who don't believe in sharing. Instead they buy a gas pump and get water illegally from the river. To be fair, there are those who have bought land because they care for the land and are indeed farming. But most are devoid of traditional knowledge.

We will also embark on a voyage that will lead us to examine the Qua'ran, the Moorish "culture of water," and the role of the Tlaxcalas. For why did Sebastián Martín, the older brother of Francisco, give part of his grant to create the Trampas grant, which was settled by the people from the Barrio de Analco in Santa Fe, a Tlaxcalan village? What is the missing link that I haven't uncovered? By 1796 the Martíns controlled all the land from present-day Alcalde to the Jicarita, all the land between the pueblos of Picurís and Ohkay Owingeh. Sebastián had the Sebastián Martín grant, which was first awarded to him in 1703. In *History of Los Luceros Ranch* Dr. Corrine P. Sze writes, "At some time before 1703, the lands along the Río Grande and north of San Juan Pueblo were claimed by a trio of men: Joseph García Jurado, Sebastián de Polonia, and Sebastián de Vargas. When they failed to occupy the grant within the required time, they lost their right to it. Thus, in 1703 the brothers Sebastián and Antonio Martín Serrano petitioned for the land for themselves, other brothers and a brother-in-law." In 1705 they finally gained possession of the lands, which extended from San Juan Pueblo north to Embudo. In 1712 the grant was revalidated. Sebastián Martín described his activities in 1712: "I have broken up lands, opened a main ditch from the Río del Norte for irrigating the land, built a house with four rooms, and two strong towers for defense against the enemy in case of an invasion, being on the Frontier," according to Dr. Sze.

The Embudo Grant was then made to Sebastián's brother Francisco in 1725. The Plaza del Embudo also had a house with four rooms and two strong towers, or *torreones*, according to the last will and testament of

Francisco Martín. Is it simply a coincidence, or was the same model used by both brothers? Up to then Francisco and his wife, Casilda Contreras, had been living on his brother's land near present-day Alcalde, then known as Nuestra Señora de la Soledad del Río Arriba, Our Lady of Solitude of the Upper River. In 1751 the Trampas grant was carved out of part of the Sebastián grant, and the Santa Bárbara grant was awarded to Valentín, the grandson of Francisco, in 1796. No wonder there are so many Martíns and Martinezes in this area; just look at the phone book. There probably isn't a family in the area that, if they go back one or two generations, doesn't have Martín(ez) lineage. I am related to the Martín Serrano clan on both sides of my family, on both my mother's and my father's sides.

Growing up, I always wanted to visit Spain, I guess because I had always heard that some of our ancestors did come from Spain—that we were descendents of the conquistadores. Though at home my dad always said we were mexicanos. Of course, when I started school in Dixon, historically known as Plaza del Embudo, I realized that the language of my parents was not the same as the language I was expected to know upon entering school. All of a sudden my world was turned upside down. My first book was Dick and Jane, and what they ate was completely different from what I was familiar with at home. They ate cereal, I ate atole and *chaquegüe*, and I still do; they ate white-bread toast, I ate tortillas. Then the teachers—who also happened to be from the community, like me— told us that our ancestors came from the east on the *Mayflower*. But when I came home the language of my parents was Spanish, the food we ate I thought was Spanish, only to later find out it was indigenous, and we had come up from Mexico, not Plymouth Rock. Something didn't make sense.

When I was in the village of Arellano, Spain, in the region of Navarra in the Basque country, in 1997, archaeologists had just uncovered some Roman ruins, including some cement pipes used for irrigation from the first century after the birth of Christ. This was part of our history, part of my history, for that area became known as Arellano in the mid-1300s, but people had lived there at least since Roman times.

It's not in the vein of nostalgia, or thinking about "the good ol' days," that I write about acequias but rather with an eye toward the future, the twenty-first century and the new millennium. It's about an ethic and aesthetic whose roots go back thousands of years, about organic gardening, building terraces, and planting by the cycles of the moon, techniques that are in vogue today and that our ancestors practiced for thousands of years, on both sides of the Atlantic. These ideas were later incorporated into the Laws of the Indies and also the Plan of Pitic of 1783 in what today is Hermosillo, Mexico.

In the search for who I am, why I am the way I am, a place that keeps popping up is La Soledad del Río Arriba, today in the vicinity of present-day Alcalde and the surrounding communities of Los Pachecos, La Villita, and Los Luceros. Why such a beautiful name, meaning "the solitude of the upper river," disappeared from the lips of the people to be replaced by such a generic name as Alcalde, or "mayor" in English, baffles me. To go from a poetic and mystical name to one that is governmental and bureaucratic is puzzling if nothing else. Los Pachecos, La Villita, Los Luceros all still go by the names with which they were originally baptized by those who first called these fertile lands home. Even the most ancient names, such as Cachanillas, undoubtedly an indigenous name; El Guique, on the west bank of the river; and Estaca still persist today.

When I first started working at the Oñate Center above La Villita, Alfredo "Bully" Trujillo, a longtime friend who recently passed away, told me, "This is your home; the Arellanos are from here; the Reeves property used to belong to the Arellanos, my grandfather used to say." Then as I started doing research I found out that indeed the Arellanos had lived there at one time. And not only that, this was the homeland of the Martín Serrano clan, one of the original families that had come north with don Juan de Oñate in 1598. And after retreating to El Paso del Norte, present-day Juarez, during the Pueblo Revolt of 1680, they were one of the original families that returned with don Diego de Vargas after 1692 to reclaim their ancestral homeland, where today's Hacienda de los Luceros stands.

But it is the Camino Real de Tierra Adentro, a name I first heard as a child growing up in the Embudo Valley, a name far more captivating and romantic than simply Highway 68, as it is now known, that brought the world to us and also connected us to the outside world. It is still the same space that connects me to my past as well as to my future. For my history is not from east to west but rather from south (Zacatecas and Aguas Calientes) to north or vice versa. Today La Villa de Albuquerque is the heart of that cross, the Big I is that which connects us in all directions. In a way the Camino Real, or Royal Road, and the Santa Fe Trail intersect here, though the original trail ended in Santa Fe.

How is it then that neocons such as Lou Dobbs and Pat Buchanan want to sever this tie that has connected my ancestors for thousands of years and erect a wall? It's like cutting a body in half, severing the head from its feet. Little do they realize that the mighty Río Grande in 2006, for the first time in my memory, refused to die and fought back mightily, with all its tributaries overflowing its banks in joy. And the farmers, especially those who had only recently moved here, instead of crying about drought replaced the dreaded word with *flooding*. The year 2006 will go down in history as one that started with everything burned due to what alarmists were calling a drought, then turned into a bumper crop of water.

My mother would always say we need to *siempre tener fé*, or "to have faith," as she would *matear*, "sow her seeds." And how did she prove her faith in the Almighty? Every spring, whether there was snow in the Jicarita or not, she would go out and with her calloused hands like the dirt in our farm, soft yet hard, she would bend her back and deposit the tiny chile and corn seeds, among others, in the *carreritas*, rows she would patiently make with her hoe. A hoe that after more than half a century in use knew how to massage the soil so it would produce. And produce it did.

Water was always on the lips of my parents: not enough water, just the right amount of water, or too much water when the arroyos ran and caused damage to the acequias or the fields. My parents talked about water when planting chile, corn, calabazas, and cucumbers. For without water there would be no crops or fruits. We also had all kinds of fruits: *manzanas mexicanas* (Mexican apples), *cerezos mexicanos* (Mexican cherries); everything was related to Mexico, though I had never been to Mexico. But I was Mexican because my dad always said,

Nosotros somos mexicanos, "We are Mexicans." And those who spoke only English were *gabachos* or *gringos*. I never heard the word *Anglos* until I was much older. Later, when I went to Spain in 1998, while I was waiting for the archives to open in Lorca, I found myself talking to a gentleman sitting in his patio, who said nonchalantly, *Los gabachos son muy malos*. It struck me like a bullet in the middle of my forehead. "Gabachos," I then learned, were to him the French, but to us nuevo-mexicanos the word referred to the Anglo-Saxons, or *los Americanos*. I mention the above vignette because I never realized when I was growing up that what my parents called *huerta* was a Latin word, or that *acequia* was Arabic in origin and *milpa* was from the Nahuatl, from the Aztecs in Mesoamerica. They also used the word *labor* to refer to what was growing in the garden. They would say, "Que bonita está la labor" (The garden is beautiful), or "Este año no está muy buena la labor" (The garden is not doing very good this year). Three words that I grew up hearing almost on a daily basis—*huerta, milpa, labor*—but I never questioned their origin; they were all part of being a "mexicano," of Mexican origin. But when you saw a beautiful weaving or some other some other creative work, they would say, "Que bonita labor" (What a beautiful work).

I want this book to read like a novel, a novel of knowledge and wisdom—*de sabiduría y juicio*—of perseverance, resilience, and stubbornness; about humankind's timeless work upon the earth in order to draw sustenance and life. This is in no way a scholarly work but rather one in which traditional knowledge is the main thread that ties the work. What better example of surviving in a harsh environment amid climate change than the collection and conservation of water and the growing of food in a sustainable way? This is a work, then, not so much about new innovative ideas but rather about looking back with a critical eye and listening to those whose footprints we now follow, to the murmur of their words that echo in the distance. This is about sharing knowledge in the form of a story; farmers know that you can't experience what you don't plant. "Using traditional knowledge does not mean to reapply directly the techniques of the past, but rather to understand the logic of this model of knowledge"; this according to

Pietro Laureano, writing in *Ancient Water Catchment Techniques for Proper Management of Mediterranean Ecosystems.*

Let us now embark on a journey that will begin in the Middle East, where the acequias seem to have originated, and from there follow the *caminos de agua*, the water roads, that will eventually lead to Embudo, to my al-munyah, an experimental orchard-garden and recreational site at *la junta de los ríos Embudo y del Norte*, today known as the Río Grande.

Tunelito at Valle Allende, Chihuahua. Photograph by the author.

Sacred Water

Origin of Life, Drink of Knowledge

≈

El duende del agua, la llorona loca bailando en el bordo de la acequia.

EXPLORING ARID LANDSCAPES
AROUND THE WORLD

The Middle East: The Yemen Connection

THE PRECEDING SECTION WAS MORE autobiographical, and in this section we embark on a journey to look at other community open-air irrigation systems, all of which in one way or another relate to the acequia systems introduced by the Moors into Spain after 711; with Cortes, this system became one with what existed previously. This hybrid mestizo system is what eventually made its way to New Mexico in 1598, following the Camino Real de Tierra Adentro, also referred to as the Camino de Agua. This in no way negates the fact that the Native Americans of the American Southwest were also irrigating prior to this date. But that's another epic journey to still be traveled.

The word *acequia*, based on the research I've done, seems to have its roots in Yemen. A *saqiya* was the cupbearer of water or wine, a courtly and poetic symbol. *Saqiya aur pila*, "Cupbearer, serve me more." Sabaean was the language spoken by the Yemenis before Islam, and it seems most of the words related to hydrology came from that language. Here, then, is where our global acequia sojourn begins. The scenery in the Harraz Mountains as seen in photos is breathtaking: cultivated terraces rolling down fertile slopes, with the backdrop of jagged mountains common to all desert environments. On the ridges, villages cling to the peaks.

On the carefully constructed terraces, coffee plantations flourish. Here agriculture is practiced more intensively than in other parts of north Yemen. The area is known as the Fertile Mountains because it benefits from bountiful monsoon rains. The terracing, carried out in such a fine and impressive manner, has been so carefully maintained by farmers that the terraces have survived for thousands of years. In addition to coffee, millet, rye, wheat, barley, lentils, and beans have been grown on these multiterraced fields for centuries.

Since the word *acequia* seems to have been born in this type of environment, our journey to understand water and community begins here, for here we see a place where people definitely had the knowledge of the water and the wisdom of the land. Only people with such knowledge and wisdom can survive in this harsh environment. As we embark on our journey we will see other God-forbidden environments where humans have not only survived but thrived and in the process have found the knowledge and wisdom embedded in the landscape, which became their greatest teacher. In the Middle East, Asia, and North Africa there are also underground acequias, called *qanats*; in Mexico they are called *foggeras*.

An article in *Science Daily* of July 21, 2008, reports that archaeologists in Yemen found traces of irrigation systems indicating a transition from herding to farming 5,200 years ago. According to Michael Harrower, Department of Anthropology, University of Toronto, "Agriculture in Yemen appeared relatively late in comparison with other areas of the Middle East, where farming first developed near the end of the last ice age about 12,000 years ago. . . . It's clear early farmers in Yemen faced unique environmental and social opportunities and challenges. Our findings show farming in southern Yemen required runoff diversion technologies that were adapted to harness monsoon (summer) runoff from the rugged terrain along with new understandings of social landscapes and rights to scarce water resources."

A chronology that accompanies the article "Buscando Nuestros Antepasados Benidorm, Noria . . . Vienen de Yemen y Otras Muchas" in the daily *El Mundo* of February 16, 2004, gives us an idea as to the etymology of words familiar to those of us studying traditional irrigation and farming. It says that each Yemeni warrior in the Omeya army was accompanied by ten civilians, among them family members and experts in irrigation who carried with them seeds and tools needed to work the land. This was the same model used by Juan de Oñate when he marched north from Zacatecas to Ohkay Owingeh in 1596, accompanied by four hundred families of Indians from Mesoamerica. They served the same function, since they were the farmers. Though the Yemeni warriors brought war, they also brought with them revolutionary systems of irrigation, as well as rice, peaches, and oranges.

The Yemeni warriors also brought a language, Sabaean, to name places and other things in the peninsula. Many words with Arabic roots actually come from the language of the descendants of the queen of Saba. Words from Sabaean related to water include *alberca* (*al-birka*), "cistern for irrigation"; *acequia* (*assaqiya*), "irrigation canal"; *zanja* (*az-zanija*), "channel that is sculpted in rock"; and *noria* (*na'ura*), "waterwheel," an old-fashioned well in New Mexico.

According to the scholar Hadi Eckert of the Office for the Protection of Historic Cities in Yemen, the majority of the Yemenis in Spain from Saná were campesinos from the mountains of eastern Yemen. Even though Arabic was becoming more common, Sabaean continued to be used to name the flora and fauna and for all the vocabulary for irrigation and agriculture. That still continues today. Sabaeano is a Semitic language that flourished about three thousand years ago near the civilization of southern Arabia, which was centered in Mareb and predominated in what today is Yemen. In the year 628 the Sabaeans became part of Islam and adopted the Arab language from the Qua'ran. This Arabic contaminated with Sabaeano or vice versa was what was spoken by the Moors and Berbers when they invaded the Iberian Peninsula in 711 AD. And with this language they named towns, rivers, and mountains.

For cisterns, they used several concepts to differentiate one type from the other: *al-marhaw*, a cistern designed to capture water and allow it to overflow; *al-jirnana*, a cistern for the storage of water; *al-karif*, a cistern that gathers the water that comes from the mountains; and *dirwen*, a subterranean cistern sculpted on rock.

Other Sabaean words that relate to water are:

> *as-sirr* (*as-sarr* in Sabeo-Arabic), watercourse in the mountains. The word *sierra* seems derived from this word.
>
> *al-jahl*, a rapid watercourse with waterfalls.
>
> *as-sirb*, a concept relating to acequias, a person's turn to use the water for irrigation.
>
> *al-jisr*, a diversion dam with steps to slow down the water.
>
> *al-aqm* and *al-maqam*, partial diversion dams that direct the flow of water or establish the volume amount.
>
> *an-nahr*, channelized torrents of water.
>
> *wad*, a permanent watercourse such as a river. The word *Guadalupe*, (the "lobos river") also has its roots in *wad*.

In addition, there is the Guadalquiver River in Andalusia. *Guadalquivir* means "big river," such as the Río Grande.

Much of the vocabulary used today for irrigation, agriculture, and construction, the central elements of the civilization in southern Arabia, is encountered in the Sabaean language. The majority of extant Sabaean inscriptions are related to a hydrology project, palace, or temple financed by some king. This has permitted scholars to detect the Sabaean origin of words spoken in Arabic by the Yemenis. Yemenis also made great contributions to the Spanish language, as shown above, in terms related to the storage, management, and distribution of water. Contrary to the Romans, famous for their big hydraulic projects, the Yemenis specialized in the microengineering of water for community projects.

Accustomed to having to reclaim land from the desert in order to cultivate it, the Yemenis brought to Spain their advanced hydraulic knowledge, and from there it spread around the globe. The origins of the names that came to signify canal, torrent, river channel, and waterwheel in many cases are from Sabaean.

According to the Global Acequias Trail website conceived by Dutch sculptor Ida Kleiterp, acequias spread from Yemen throughout North Africa, then to the Iberian Peninsula, and from there across the ocean to the Americas. The "extraordinary architectural forms that she saw in the mountainous landscape of the Alpujarras" during a visit to Spain in 2001 inspired Kleiterp's sculptures of acequias. These acequias are a "network of reservoirs, channels, and movable gates that together form a remarkable ancient irrigation system." I had the opportunity to visit the Alpujarras in 1998 and then again in 2004, and what a marvel these villages are.

The Desert and the Oasis Culture

Now let's move from Yemen to the use of the land and water. Through the use of suertes, strips of irrigated land used for gardens and orchards that were granted to settlers outright (versus the *dehesas*, which were part of the common lands and usually located away from plazas), the *paisano* of northern New Mexico was forced to learn water-conservation practices as part of clearing the land for cultivation. The best model to use for the Río Arriba bioregion is the land grant division of land, where the commons (*sierras, montes, dehesas*) act as the storage or collector of the water and the acequia becomes the distributor, while the suertes (*altitos, jollas*, and *vegas*) are the benefactors or users, and the *ciénagas, bosques*, and *esteros* are the purifiers of the water. This philosophy of water conservation can be attributed mainly to the Moors, who were in Spain from 711 to 1492 and who, through their culture of water, had a profound influence on the Iberian Peninsula and its agriculture. But this same knowledge and wisdom were already practiced by the Incas in Peru and in Mesoamerica, as well as in what today is the American Southwest.

This work, then, is an overview of how diverse systems of traditional knowledge from around the globe relating to land and water use are adaptable everywhere there is a desert environment. The same techniques used in the high desert of New Mexico are found in the *bisses* system of irrigation in Switzerland or in Colca Canyon in Peru.

As nuevomexicanos, mestizos, we're heirs to two of the greatest civilizations humankind has known, but sadly they haven't received their due, as Islam is now seen as a terrorist religion and the great civilizations of Mesoamerica are nothing more than footnotes in Western history. (Today's Mayas are the illegal aliens in Arizona, among other states.) Yet they are my ancestors, and I am very proud of their accomplishments when it comes to science and technology, especially hydrology and botany. When Northern Europe was still in the Dark Ages, Córdoba, in the Andalusian region of Spain, was already a metropolis of half a million people with indoor drinking water and libraries with thousands of books.

Richard Covington, a Paris-based writer, writes in the *Saudi Aramco World* magazine, "From the ninth through the 16th centuries, Islamic societies from Spain to Oman experienced a 'golden age' of science and technology. One of the most important of these technologies is also today one of the least thoroughly studied: hydrology, or the control of the movement of water." He continues, "Integrating, adapting and refining irrigation technique[s] and water distribution methods from India, Asia and Rome, Muslim water engineers, starting as early as the seventh century in Arab countries and around the 10th century in Spain, built an agricultural revolution."

Mohammad El Faiz, an economics historian at Cedimes University in Morocco, wrote in his essay "Horticultural Changes and Political Upheavals in Middle-Age Andalusia," in *Botanical Progress, Horticultural Innovations and Cultural Changes*, "During recent years my own research on the medieval Islamic economy uncovered evidence— behind the massive transformation of agricultural activities—for a true 'agricultural revolution.'" Even though this revolution, also noted by Dr. Idrisi Zohor in his paper "The Muslim Agricultural Revolution and Its Influence on Europe," began in the Muslim Middle East between the

Irrigated garden, Egypt. Photograph by Ellen Miller.

eighth and tenth centuries, it didn't come to fruition until two centuries later in Al-Andalus, as the Arabs called southern Spain. Zohor writes, "Sevilla . . . Córdoba and Toledo became an agricultural capital and the Mekka of agronomists." He writes that Islam produced the first "tropicalization" of Mediterranean Andalusia, before the encounter with America produced a second, larger "tropicalization" beginning in the sixteenth century.

While medical and botanical books date back to the earliest civilizations, records from Egypt, Mesopotamia, China, and India reveal a tradition that predates the advent of writing. In the West, nothing suggests such antiquity. Yet in Andalusia, from the tenth to the fourteenth centuries, countless treatises on agriculture, botany, and medicine were produced by the Hispano-Arabs who called the Iberian Peninsula home. In the tenth century a Persian mathematician, Muhammad al-Karaji, wrote "The Extraction of Hidden Waters to the Surface" in Bagdad. Ibn

Sina (980–1037), known to most as Avicenna, established hydraulics as an independent discipline. He was a physician-philosopher whose discoveries about water had far-reaching impacts on agriculture, science, law, and social organizations. Not only Spain and northern Africa, but eventually lands across the Atlantic, including the southwestern United States, were affected. After the sixteenth century Arab-inspired technologies were adopted not only in the Canary Islands but also in what is now New Mexico, Texas, and Louisiana, according to Dr. Antonio Malpica Cuello, a professor of medieval history at the University of Granada.

Malpica Cuello and Dr. Carmen Trillo San José, colleagues at the University of Granada, have written extensively about Arabic hydrology in Spain. Trillo San José's book *Agua, tierra y hombres en Al-Andalus* is a compilation of all her previous work on hydrology and Nazarine agriculture in Granada and the Alpujarras. The work of Malpica Cuello and Trillo San José, more than any other, opened my eyes to the expansion of Arabic agricultural techniques to wherever the Spanish Empire expanded, including the Philippines.

In 1997 I walked through the narrow streets of Sevilla in search of a copy of *Obra de Agricultura*, written in 1513 by Gabriel Alonso de Herrera, which I had been unable to locate in Madrid. As I walked out of a bookstore close to the hotel where I was staying, a book titled *Libro de Agricultura* by Ibn Bassal caught the corner of my eye. Immediately I backpedaled and went inside and asked for the book, the last copy they had. I bought it immediately. When I opened the book back at the hotel while enjoying a glass of fine Spanish wine, I noticed it had been written in 1075. I couldn't believe my eyes, but what surprised me more once I started reading it was how much it seemed to be talking about the agriculture I was familiar with in northern New Mexico. As I started doing more research, I realized that between the tenth and fourteen centuries, what some scholars call the "Agronomy School of Sevilla" had produced a plethora of works on agriculture. Only recently have they been translated into Spanish, but none yet into English. Later, through friends in Spain, I was able to acquire other titles. These books, then, introduced me to the work done by scholars such as Malpica Cuello, Trillo San José,

Dr. Expiración Garcia Sánchez of the La Escuela de Estudios Árabes, and Dr. Julia María Carabaza Bravo from the University of Sevilla. These are scholars that I have also communicated with via the Internet and who have shared very valuable information. Then there are also Dr. Thomas Glick from the University of Boston, the premier historian on Muslim irrigation practices, and Dr. Luis Pablo Martinez from Murcia, who have also shared their knowledge with me and other nuevomexicanos such as Dr. José Rivera of the University of New Mexico, author of *Acequia Culture: Water, Land, and Community in the Southwest.*

In terms of city building in the New World, both the Mesoamerican and the Muslim influences have largely remained unacknowledged. Not much seems to have been researched in terms of the influence of Islamic city patterns—the medinas—and municipal law on Spanish urban thought and the Laws of the Indies. I have yet to read anywhere a comparison of the alquerías and mercedes (land grants), to see if they have the same roots in terms of settlement patterns. To me they appear very similar.

There are at least ten treatises on agriculture in Andalusia published between the tenth and mid-fourteenth centuries that have survived, with four written between 1074 and 1110. One of the best known is by Ibn Luyun al-Tuyibi, the *Tratado de agricultura*, the "book on the beginning of beauty and the end of knowledge which deals with the fundamentals of the art of agriculture," published in 1348. It is a book I found in one of my sojourns through Al-Andalus, the name the Arabs gave to their beloved adopted land. And acequias were the system that made possible the movement of water to irrigate at times marginal land for agricultural purposes by giving life to *tierras muertas* (dead lands), as Dr. Carmen Trillo San José writes. The Moors seem to have been more concerned with the aesthetics of the land than were the Christians, as indicated by the literature on this subject.

A news item from Reuters, dated August 23, 2006, reported,

Archaeologists in Israel have unearthed an ancient water system which was modified by the conquering Persians to turn the desert into a paradise.

The network of reservoirs, drain pipes and underground tunnels served one of the grandest palaces in the biblical kingdom of Judea.

The palace, first discovered in 1954, where the communal Ramat Rachel farm now stands, recently unearthed nearly 750 square feet of a unique water system. But it was the Persians, who took control of the region around 539 BC from the Babylonians, who renovated the water system and turned it into a thing of beauty. Oded Lipschits, a Tel Aviv University archaeologist, said they added small waterfalls to try to turn a desert into a paradise.

"Imagine on this land plants and water rushing and streaming here," Lipschits wrote. "This was important to someone who finds aesthetics important, for someone who wanted to feel as though they are not just in some remote corner in the desert."

For as desert people in the Río Arriba bioregion, we are also descendants of the Nabateans who settled the Negev Desert of Israel and whose elaborate water systems made the desert bloom. Sadly, we have forgotten this knowledge, because the schools don't think it's important; they are more concerned with turning everyone into global consumers than with teaching the wisdom of the land, of the land where we live, what I refer to as our querencia.

How ironic that those societies that have evolved from the deserts or semiarid places are the ones that have developed a culture based on water. It's when water is not available that it's most valued and appreciated. On a couple of occasions when our hand-dug well has gone dry, we've realized how much water we use—to drink, cook, clean, for the animals; life cannot exist without water. *Sin agua no hay vida*, "Without water there is no life." When the faucet goes dry, all of a sudden I become thirsty. For that reason those cultures that have evolved from the desert have developed very sophisticated networks, not only for irrigation but also for extracting water from the earth or harvesting it from the sky.

Scholars cannot pinpoint exactly when or where the practice of irrigation started, and they probably never will, but some have traced what is believed to be the beginnings of agriculture to Jericho, to 8,000 BC. Marvin J. Wolf, in *Ambassador,* described water in Jericho:

> "Look down there," says Eduard, our guide. "See how each layer has a slightly different texture and color, a distinctive mix of soil and stone. . . . Near the bottom is a layer of light rock. Probably the foundation of a house built around 8,000 B.C. Don't forget, you're in Jericho, the oldest city in the world."
>
> Jericho, which some historians insist on calling a "settlement" rather than the more grandiose "city," flourished throughout antiquity because it had a reliable source of water—the copious spring of Elisha mentioned in Second Kings, Chapter 2:19–22. The original settlers, Stone Agers that archaeologists call Natufians, dug ditches from the spring to irrigate their crops. These were probably the first people in the world to make the transition from nomadic hunting and gathering to agriculture.

Biblical Background

From 2 Kings, 2:19–22:

> The men of the city said to Elisha, "Look, our lord, this town is well situated, as you can see, but the water is bad and the land is unproductive."
>
> "Bring me a new bowl," he said, "and put salt in it." So they brought it to him.
>
> Then he went out to the spring and threw the salt into it, saying, "This is what the LORD says: 'I have healed this water. Never again will it cause death or make the land unproductive.'"
>
> And the water has remained wholesome to this day, according to the word Elisha had spoken.

So the water of Jericho has been wholesome to this day, according to the word that Elisha spoke. I have dreamed of visiting the Middle

Field system, Egypt. Photograph by Ellen Miller.

East and Jericho, which is described as a beautiful, green oasis. The "Date City," as it is called, is thought to be the oldest continuously inhabited city in the world—dating back nearly ten thousand years. It is located beyond the northern portion of the Dead Sea (the lowest place on earth). Jericho is also the lowest city on earth—some eight hundred feet below sea level. Nearby are the mountains (between Jericho and Jerusalem) that capture water far above Jericho and then carry it through underground springs to where it spews out in this desert oasis.

One of these springs is Elisha's Spring. Some scholars claim that the first acequias were those that had their origins from this spring, since they were only about fifty yards from its exit.

Acequias, as we know them today, with roots in the Middle East, appear to have evolved at least eight thousand years ago, if not in Jericho then in the Indus Valley. Archaeologists also have recently uncovered ancient irrigation systems in what today is Peru. These date back at least to 6,000 BC, which means that irrigation came into being around the same time on different continents. Native American tribes of the

Southwest and in Mesoamerica had also developed sophisticated irrigation systems prior to the arrival of the Spanish in 1492.

In Ladakh, India, acequias are known as *yuras*; the *presa*, or dam, is known as a *raks*, while in the Philippines they are called zanjas, a word also familiar in New Mexico. The person who cares for the water is known as a *mayordomo* in New Mexico; in Chihuahua, he is called an *aguador*. In Tucson, Arizona, the ones who takes care of the water is also known as the *juez de agua*, the same as in Valencia, Spain; there he is also referred to as *cequier*. In Ladakh he is the *chud-pon*, different name but the same responsibility. *Chud-pon* is derived from *chu*, meaning water, and *pon*, the appropriator of water.

Traditional Water-Harvesting Techniques:
Terraces, *Albercas, Aljibes,* and *Cajetes*

Terracing is a way to grow crops on steep, erosion-prone slopes, and it is an efficient method of water conservation. It is a technique that has been practiced around the world since humans started doing agriculture. Terracing seems to have developed around the same time as irrigation in arid parts of the world, whether in Asia, Sri Lanka, Lebanon, Yemen, Morocco, the Iberian Peninsula, the Andes, or Mesoamerica. The *cajete* terrace system, for example, has been in use since pre-Hispanic times (1000 BC) in Tlaxcala. Irrigation and terracing seem to be part of the same system of agriculture. But like the acequias, terraces are in a state of disrepair almost everywhere. Terraces are as much a method of water harvesting as are cisterns and *aljibes*. People in arid lands have always been very ingenious when it comes to conserving water, whether it is through underground canals known as qanats, or foggeras in Mexico, or the cajete terraces used by the Tlaxcaltecas. Yet this traditional knowledge is rapidly disappearing.

Today water harvesting is illegal in the state of Colorado, which doesn't make sense, since a lot of the surface water, especially during heavy rainstorms, ends up as floodwater. Thus instead of it being used beneficially it becomes a problem, flooding homes and roads and silting up acequias. Everywhere that water is scarce, people have depended on cisterns and aljibes to store water, as well as albercas to store water for irrigating small gardens. Cisterns and aljibes are another element incorporated into a semiarid landscape for water sustainability. They are part of the knowledge of the water and the wisdom of the land. J. Labasse, in *La organización del espacio*, wrote that water is "the most ancient and most spectacular objective of the organization of space to assure the inhabitant and the producer regular use [of water] independent of floodwater." The oldest and most elemental form for this artificial storing of water, constructed during the Roman period, was simply a hole in the ground to gather storm water or a small earthwork to retain the water. These small dams were called *lacus*; later they came to be called *piscina* when they were constructed in a geometric form and *cisterna* if they were covered.

Both *alberca* and *aljibe* come from Arabic. The word *aljibe*, from *al-yubb*, means a container that is dug out totally or at least partially, where water is stored and which is usually covered with an arch, or *bóveda*. Ramón Germinal gives the origin of the word as *algúbben*, an Arabic word from a dialect spoken in Andalusia. This type of storage system has a long tradition in the Middle East, where cisterns are known as *harabe*, meaning "ruins"; the Hebrew term, *maagurah*, has been known since biblical times. In the Iberian Peninsula, according to scholars, cisterns dating back as far as the Bronze Age have been found around Murcia and there are some that also go back to the Roman period, but most of them are from the Muslim era. During this period the Arabs would build aljibes along roads so that water would be available to travelers. Water was conducted to these aljibes via *boqueras*, a sluice in an irrigation canal; *agueras*, a trench for irrigation; or acequias. There are three types of aljibes: those that are shaped like a bottle and are known as *jarra* or *botella*; those that resemble a well and are called a *pozo* or *xeringa*; and those that are shaped like a cistern, or *bassa*. These are usually used for watering livestock and are also known as *abrevaderos*. Some aljibes even have fish or eels, as they move the water so that it doesn't stagnate.

In Petra, Jordan, they have a device for harvesting humidity along a slope, called a *khottara*, that works by means of channels that catch water on the slope and convey it to pools underneath.

Terraces and man-made irrigation systems are found throughout the world. From Afghanistan to the Alpujarras to the Philippines to Peru, the world's terraces and the man-made artificial water systems that supply them are superb examples of sustainable land and water use, allowing a high-density population to live in what appears to be an inhospitable terrain. These scenic and exotic works are the product of massive organized human efforts. Each terrace is irrigated by water that is transported down the mountainside from springs, rivers, or reservoirs using a complex system of earthen canals. It was American historian Karl Wittfogel who first called these complex societies "hydraulic societies." Today many of these terraces are abandoned as people work for wages instead of maintaining such highly intensive agricultural systems. But when these systems are not maintained, erosion and flooding become a problem.

EUROPE

The *Bisses* of Valais: The Acequias in the Alps

In the course of researching this project, I was surprised to learn that similar water systems exist in parts of the world where I never would have thought they were possible. For example, I never thought that I would encounter man-made watercourses in Switzerland, where these canals are called *bisses de Saxon.* In the French-speaking area they are known as *rayes,* and as *Suonens* in the German-speaking communities. As in other parts of the world where this type of irrigation system exists, the channels tend to be small, normally from two to six miles long. And like the acequias madres, these bisses provide water to secondary and tertiary sources. They also transport water to small artificial lakes where it is stored for later use, the same as with the albercas found throughout Andalusia and in other parts of the Spanish-speaking world. And just as in New Mexico, wood channels were constructed along cliffs to deliver water. In northern New Mexico, a world apart, similar devices are called *canovas.*

Common-property corporations called *consortages* manage these systems. According to Emmanuel Reynard, "Hill irrigation in the Swiss

Alberca, Alpujarras, Spain. Photograph by the author.

Alps has a long tradition of more than 700 years, especially in the Rhone Valley (the Canton of Valais) where a network of more than 1400 km of channels was created. Until the 19th century, irrigation was generally limited to the meadows. In the 19th century, it was extended to vineyards and orchards." The green landscape therefore is not the result of natural conditions but of a wise use of bisses. As in North Africa and in Spain, the system creates a particular kind of landscape where settlements are located following the pattern of the bisse canals.

According to Reynard, "In Switzerland, water rights are currently regulated by the Swiss Civil Code (SCC) in use since 1912. Property rights on water are based on two principles: the principle of accession, which considers that underground property is linked to soil property (springs and underground water property are therefore linked to soil property), and the principle of state sovereignty, which restricts private property for predominant public interest." Bisses are managed through a consortage, a system of common-property rights. Consortage members are common owners of the infrastructure, and they benefit from rights to the use of the water resource or the products of the association. Diversion and allocation of irrigation water are normally organized into a cycle known as *tour d'eau*. The consortage association's members manage the canals in much the same way as the acequia commissions do. Their workdays are referred to as *corvées*, and the members are known as *consorts*. Those who care for the waterways remove rocks or other foreign objects, do repairs, and ensure that people adhere to the complex rules governing the control, use, and maintenance of the bisses.

But as in other arid areas of the globe, a decrease in agricultural production and urban expansion have led to the emergence of new water uses and water users. As in most areas with decreased irrigation usage, local people are discovering the historical and cultural value of these sites. In New Mexico the acequias were once the roads of the communities. Since about thirty years ago the bisses have started to be used for hiking and have once again become part of offical walking paths under the control of local municipalities. Instead of only being seen as pure agricultural infrastructure, bisses have now become multifunctional, serving as an interface between agriculture, culture, and tourism.

Les Hortillonnages: Marshland Agriculture

Human ingenuity is not restricted to one culture; rather, it seems that similar systems have sprouted throughout the world, almost at the same time, but with no contact among the different cultures. This can be said of the man-made irrigation systems throughout the arid world or the systems for drying up marshes and wetlands and turning them into productive agricultural lands, whether in the Andes, Mesoamerica, or the Mediterranean in Europe.

In Amiens, France, there is a unique area right next to the Amiens Cathedral, known as Les Hortillonnages. The name is derived from *hortus*, "garden," and *oire*, "small surface." It's an area of small gardens on small islands in the marshland along the Somme River, surrounded by a grid network of man-made canals, the *rieux*, which are navigable by flat-bottomed boats known as *barques à cornets*. Les Hortillonnages dates back about two thousand years to Gallo-Roman times; at that time these gardens were used to feed soldiers. Today there are only about 740 acres under cultivation. At one time there were over one thousand market gardens. Today there are just a dozen or so remaining, used for growing fruit and vegetables, but the market still takes place every Saturday morning on the Quai Parmentier in Amiens.

The best description of Les Hortillonnages comes from the preface to *Northern Europe: An Environmental History*, by Tamara L. Whited, Jens I. Engels, Richard C. Hoffmann, Hilde Ibsen, and Wybren Verstegen. Whited writes, "Just behind the old neighborhood of Saint-Leu in the northern French city of Amiens, a capillary network of canals shapes a watery enclave within view of the mighty Norte Dame of Amiens, the city's towering gothic cathedral." The gardens are older than the cathedral itself, even the original one that dates to AD 850.

The preface continues,

> Seemingly a world apart from the human activity of a prospering twenty-first century city, the canals feed the Babylonian luxuriance of the "floating" gardens bordering them. The visitor may feel the brush of an overhanging willow branch against a cheek as the narrow boat, provided by a conservation association and equipped with silent, non-polluting electric motor, wends its way through a small part of the labyrinth open to tourists. A few paces away from rush-hour traffic, warblers, wrens, nightingales, woodpeckers, thrushes, kingfishers—in all, some sixty species of nesting birds—find habitat among the rushes and reed, club moss, chickweed, and water lilies that thrive in and near the water.

The canals and parcels in the historic heart of Amiens were at one time used as an area for cultivation to feed Roman troops. Some scholars think that humans might have occupied the marshland formed by the Somme River and its tributary earlier and subsisted on the available resources. And it was years after the Romans left that medieval farmers established Les Hortillonnages by draining and canalizing the marshland. By transforming the area into cultivable plots, separated by canals, they made it the "economic nerve center of preindustrial urbanization" in Amiens. As a result, besides watering the vegetables that fed the city, the canals also provided transportation and water for the dyers who set up along the banks. Fuel from decaying plant life was another byproduct.

In 1825 the Somme was channeled, though a few branches of the river were retained to feed the smaller canals. Then, in the 1840s, with the coming of the railroad, the community of Longueau (Long Water) was further separated from Les Hortillonnages to the southeast, and in the twentieth century a new boulevard created another separation.

Like the *chinampas* of Xochimilco, Les Hortillonnages was becoming an obstacle to urban development. According to Whited, "Many of the vegetable plots gradually gave way to suburban idylls, complete with lawns, geraniums, and garden gnomes." In 1975 Nisso Pélossoí, of Greek origin, mobilized support to stop a projected road that would have destroyed the gardens. Finally, in 1991 France's Ministry of the Environment listed them among its Paysages de Reconquete, or "reconquered landscapes." According to Whited, "reconquest" implied "the paradoxical preservation of a landscape from its two creators—water and humanity." Like the acequias, Les Hortillonnages is an artificial, man-made landscape; it is unique not only to France but to the world.

With the omnipresence of water and the ability of people to manipulate the environment into small-scale agriculture, not only do Les Hortillonnages echo the environmental history of northern Europe, but they also produced up to three crops per year.

Wherever there is land that has too much water, a ciénaga, or marshland, develops. Those who have lived in such an ecosystem have always found a way to cultivate food and to survive, to thrive, turning what would at first appear as a hindrance into fertile "floating gardens." Such gardens were developed in various parts of the globe by people who were basically discovering the same hydrological techniques for using the excess water, *las escorrentinas*, and manipulating it so that it could be used to grow food. These techniques for turning dead water into live water sprouted in Amiens, France; at Xochimilco in Mexico City; on

Ibiza, an island east of Spain in the middle of the Mediterranean; and in the Andes. They were used under the Romans, Aztecs, Arabs, and the Aymaras and Quechuas. In France these "floating gardens" are called Les Hortillonnages and in Ibiza, Ses Feixes, an Arabic name; the ridges or mounds are known in Nahuatl as chinampas and in the Andes as *camellones* in Spanish and *suka kollas* in Aymara. In a sense they are raised gardens, created from the muck in the bottom of the water, that allow the water to clear up and thus clean itself before being used to water gardens and produce enough food to feed the surrounding population. These gardens in the water, created from waste, became sustainable, producing up to three crops per year, which were sold at the market or used to feed a certain population. Remarkable systems for recovering a certain type of landscape that doesn't produce unless altered and that becomes a problem for settlement, these gardens convert marshland into very productive and aesthetically pleasing creations, pieces of art portrayed by nature's true colors and textures.

This type of agriculture, which is very intensive and highly productive, is called *agricultura de los humedales*, "marshland agriculture." Some writers have compared Les Hortillonnages and the chinampas, or the chinampas and the camellones, but the four areas with such an agricultural landscape and hydrology—in France, Ibiza, Mexico, and the Andes—have not been analyzed together. Could it be possible that those who developed such systems in different periods of history and in vastly different landscapes (except that they all had one similar element, which was more water than needed for traditional agriculture) have by some means communicated? Or is it that humans, as they have developed, have gone through the same thinking pattern and come up with the same solution to the problem of having more water than needed? That solution is to drain the land and rearrange the landscape. In terms of altitude, such systems range from sea level for Ses Feixes to seven thousand feet for the chinampas and up to twelve thousand feet for the camellones.

Landscapes where the acequia culture developed and those of the ciénaga, or agricultura de los humedales, are ironically part of the same arid landscape. Acequias prefer the joyas, or hollows, while the humedales prefer the ciénagas or marshlands. All such systems are

Chinampas, Xochimilco, Mexico City. Photograph by the author.

based on the communal system of cooperative labor: the concept of con-vite when it comes to food and the *repartimiento* when it comes to water. With acequias, sharing is done in times of water shortages, while the "floating gardens" can also be said to be a system for sharing water, as the water has to be maintained at a level at which it will not dry the soil or flood the raised beds.

Moorish Influence on the Iberian Peninsula

The history of acequias and agriculture can be divided into three peri-ods on the Iberian Peninsula. The first is the classical period, which encompasses the writings of Cato, Palladius, and Julio Moderato Columella, who though a Roman citizen was born on the Iberian Peninsula in what today is Cadiz, a place I fell in love with when I visited the ancient city in 2004. I even went to the place where Columella sup-posedly lived. The second period covers the work of the Spanish-Arabs from the late tenth to the mid-fourteenth centuries, and the third is the modern period, represented by Gabriel Alonso de Herrera.

Those of us who are concerned with maintaining our agricultural traditions today are facing challenges no different than those that tested Columella, who was born in the year AD 1. He wrote two major works, *De re rustica* or *De los trabajos del campo* (The twelve books of agriculture) and *Liber de arboribus* or *Libro de los árboles* (The book of trees). In his prologue to *De re rustica* Columella lamented that Rome had schools for rhetoric, geometry, music, and culinary arts but no one dedicated to the teaching or study of agriculture. The same can be said of northern New Mexico both today and during the colonial period. Nowhere can one study traditional agriculture based on the acequia system of irrigation, nor has that ever been part of public education. Columella also was concerned that the youth didn't want to work the land, that instead they preferred going to bars and partying. The same thing is happening today.

The most famous agricultural treatise from the second period is the monumental work *Libro de agricultura* (The book on agriculture), written in the 1250s by Abu Zacarías Iahia, who was part of what was known as the School of Sevilla. Abu Zacarías summed up the work of all the other writers of his time. Of a piece of land he says, "La heredad dice a su dueño: hazme ver tu sombra, cultiva" (The piece of land says to its owner: make me see your shadow, cultivate). He added, in the order in which it is said that Mohammed gave this advice, "Search for sustenance, harvesting the fruits that the earth produces." Ibn Bassal, in *Libro de agricultura* (1075), talks about irrigation using acequias, as does Ibn Luyun in *Tratado de agricultura* (1348). Some of these books have been translated from Arabic into Spanish but not into English.

The first book written in the Castilian language was *Obra de agricultura* (Treatise on agriculture), by Gabriel Alonso de Herrera, first published in 1513. In October 2006 Ancient City Press published *Ancient Agriculture: Roots and Application of Sustainable Farming*, the first English-language edition of Herrera's treatise, a compilation I worked on for ten years.

Though most scholars date the acequias in Spain to the Arab period, after AD 711, recent scholarship indicates that the Romans had earlier constructed canals that were used for irrigation. Almost all agree that the Arabs, using existing Roman infrastructure, constructed the acequias in the Iberian Peninsula. A recent archaeological dig in Monforte del Cid, according to the Alicante newspaper *Información*, November 30, 2007, found the remains of an old acequia in Agualejas, which dates to the first or second century AD.

During this period, one of the big differences between Christians and Muslims in terms of philosophy was that the Christians didn't believe in getting involved in working the land personally, in getting their hands dirty. The Muslims, by contrast, especially those who were cultured, believed that it was very important to work their land themselves. Artists, poets, writers, and statesmen all worked the land, in contrast to the Christians, who believed that only the poor and uneducated were condemned to work the fields. We are now seeing a philosophy like the Muslims' return, as we see scholars, writers, artists, and intellectuals getting involved in organic farming, sustainable agriculture, and permaculture.

Dr. Zohor Idrisi's paper "The Muslim Agricultural Revolution and its Influence on Europe," as well as the work of many other scholars, expounds on how the Iberian Peninsula flourished with never-before-seen fruits and vegetable brought from Asia, India, and the Arabian Peninsula. Possibly the biggest Moorish influence had to do with the foods that were introduced, among them asparagus, almonds, pistachios, and citrus. In addition, the culture of the table changed with the coming of the Arabs, as the Moors introduced the use of crystal and the fork. The Muslim agriculture revolution began in 711 and lasted until 1492, when another agricultural revolution began when Columbus encountered what today is known as the Americas.

The Muslims also brought revolutionary social transformations through changes in the system for ownership of land. To the Moors who settled southern Spain (Andalusia), the communal lands were known as *harim* and *mubaha*; another type of common lands was known to the Arabs as *mawat* and became known as *tierras muertas*, "dead lands," or *tierras baldías* in Spanish. All three types of lands could not be sold and were to be used in common by all the people of the alquerías, as these regions were known in Andalusia. The word *alquería* is derived from *alcarria (qarya)* and signifies *aldea,* a village or hamlet. Alcaria is a lady's name in New Mexico. Dr. Carmen Trillo San José of the Universidad de Granada recently wrote an excellent book on this topic, *Agua, tierra y hombres en Al-Andalus: La dimensión agrícola del mundo nazarí.* I have found no scholarly work comparing alquerías to land grants, but to me they appear to be very similar. Whereas an alquería seems to comprise one hamlet, a land grant sometimes comprises several hamlets, but both are surrounded by common lands.

Ibn Al Jatib, considered the most knowledgeable of Muslim historians in Granada, described the acequia as a structure "that flows on the side of the road [and] has an abundance of water, which is a marvelous occasion, with its delicious vegetable gardens and incomparable flower gardens, a benign climate, and very sweet water, on top of its panoramic and splendid vistas. It's a scenery full of myrtles; there are well-protected palaces and mosques where there are a multitude of people and high fortified buildings."

The Islamic world has always had a preoccupation with the purity of water; for humans, God created water. The concept of the purification of water, mingled with aesthetic and poetic ideas that manifest themselves in the architecture of water common throughout the Islamic world, is today very much a part of us and very much a part of our acequia culture. Water is considered a *don de Allah,* and metaphorically it then becomes *la bebida de la sabiduría,* "the drink of knowledge." For us—Indo-hispanos—water is also holy, as we refer to it as *agua bendita,* and by the same token the land is recognized as being sacred, *tierra sagrada.*

According to Cherif Abderrahman Jah and Margarita López Gómez in *El enigma del agua en Al-Andalus,* "The Moorish engineers brought with them an experience learned from Syria and Iraq which demonstrated that their principal preoccupation was irrigation and the capturing of water, which was the basis of a flowering sustained economy, primarily, in polyculture." The culture of water impregnated the social, economic, and cultural life of the Islamic world, the same as here in the Río Arriba bioregion.

Acequia, Alpujarras, Spain. Photograph by the author.

Ses Feixes in Ibiza

Located about fifty miles off the coast of Spain on the third largest of the Balearic Islands, Ses Feixes of Ibiza is a magnificent example of what Arab culture brought to the Spanish island in the Mediterranean about one thousand years ago. A relic of traditional Moorish irrigation systems, Ses Feixes features innovative agricultural methods, which were unique in the world at the time, of forming a network of watering channels based on capillary irrigation. Ses Feixes—"the plots"—is an area of wetlands with a great biodiversity of flora and fauna, which supplied water to a whole series of enclosed vegetable plots, creating areas of fertile land that are now protected because of their high ecological value. During the time of the Arabs, these plots produced fresh vegetables for the city and garnered the area fame for its agricultural products.

It is said that the walk along the paths in Ses Feixes that used to link the plots is a pleasant one. The main gates to the plots, the *portal de feixa*, are still visible, although many have fallen into a state of disrepair. Their huge white stone frames with wooden doors have made them famous for their beauty and uniqueness. They were divided into three parts, two that were used for cultivation— Prat de Vila and Ses Monks— and they were connected by a third one known as Es Prat.

Pietro Laureano, the brilliant Italian landscape architect, calls the feixes' design "an ingenious hydraulic organization." Water in the feixes was regulated by floodgates, what in New Mexico we call *compuertas*. Similar to the chinampas of Xochimilco, the feixes were connected by underground channels called *fiblas*. The canals served the twofold function of draining excess water, allowing for the collection and saving of it, and of irrigating the fields during drought seasons. This irrigation system was combined with crop-rotation systems that enabled the land to produce two harvests per year.

Up to about the 1950s the feixes were the most important farmland. They grew onions, cabbages, beets, melons, and, later, corn and peppers. But once tourism replaced agriculture as the area's main industry many of the feixes started to disappear.

Laureano, an expert on traditional knowledge and desertification, explains that "if these works were not carried out it would be a swampy area in some seasons and arid or flooded by seawater in other seasons. In this way, it is possible to carry out a self-regulating process, which allows practicing intensive cultivation of both marshlands and arid lands. Open canals are about one-meter deep and flow at a lower level than the plots of land, thus keeping them dry."

He continues, "The land excavated for building the canals is used to raise the level of the cultivated land. During hot seasons when the land undergoes high evaporation, the plots absorb the necessary quantity of moisture directly from the subsoil and from the walls of the canals by osmosis and capillarity. The process is then fostered by further underground canalisations excavated in the plots. These underground canals are built with porous stones and pine-tree branches covered with a layer of Posidonia algae collected along the coast." The system he describes here is almost identical to the chinampas of Xochimilco and Les Hortillonnages in France: three very different parts of the world, all

Photograph of the model in the Ses Feixes visitor center near Ibiza Town, courtesy Mike Deeds. Photograph by Ferran Nogués.

with similar irrigation systems developed by different cultures with no contact with each other.

"This method," writes Laureano, "ensures the good running of water through piping and at the same time it allows it to obtain a certain level of permeability, in order to give the land the quantity of water necessary to keep it humid. Therefore, the irrigation is carried out from the subsoil directly to the plant roots. This technique enables [users] to save water that would be lost because of evaporation by using open irrigation methods."

SOUTH AMERICA

Irrigation and Land Use in the Inca Empire

The Valle Sagrado de los Incas, "Sacred Valley of the Incas," is located in the Andes Mountains of southeastern Peru. Part of the region runs

Cusco, Pisac, Peru. Photograph by Mylene d'Auriol Stoessel.

along the Vilcanota-Urubamba River through the Urubamba Valley, with the entire area stretching from Ollantaytambo to Huanca. The Inca Empire, in Quechuan called Tahuantinsuyo, stretched from today's Colombia through Ecuador, Peru, and Bolivia, down into Chile and Argentina. One of the world's largest empires and the biggest in the Nuevo Mundo, as the Spanish called it, the Inca territory was immense. The Sacred Valley includes places like Písac, Ollantaytambo, the Moray terraces, Yucay, and Calca. But it has to be understood that the region does not have well-defined boundaries.

The Colca Valley, situated in the midwestern part of Caylloma Province in the Department of Arequipa, ranges in altitude from 7,000 to 14,500 feet above sea level, the limit for livestock farming. Large mountains dominate the valley, characterized by a deep canyon, more than sixty miles in length, carved by the Colca River. The canyon is the deepest in the world, twice as deep as the Grand Canyon in Arizona.

The Incas were not only great builders, but they were also remarkable cultivators. Because the Colca region lacked *llanos*, or level spaces for planting, the Incas created stepped, terraced fields on the sides of mountains and hills, and they irrigated with the use of acequias or aqueducts carved into rocks. Some of these ancient irrigation systems are still in use, helping to produce food for local consumption.

Terraces are always situated in slopes and are constructed by using stone walls or earth. Irrigation channels, or acequias, distribute water to feed the terraced land; the water may come from melting snow in the rivers or from perennial and intermittent streams, springs, or lakes. Terraces, some constructed hundreds of years ago, are as effective today as any modern solution and they are found throughout the Americas and also in other parts of the world. Terraces had two purposes: to grow food and to prevent erosion of the land. In places with no terraces, over time the soil can flow down the valley due to rainstorms and cause more dramatic events such as earthquakes.

El Valle Sagrado de los Incas is Peru's most productive agricultural area. In precolonial times the land was used primarily for producing crops, mostly potatoes. The residents not only knew how to utilize the water but, like all good agriculturalists, they learned to use different types of soil for the various species they planted. At that time the Sacred Valley was one of the most agriculturally productive zones, even on a global scale. While most Europeans were waiting for rain, the Incas, like the Moors in the Iberian Peninsula, were irrigating their land. This eliminated the threat of drought, at least where there were rivers from which water could be diverted to ensure life for the plants.

The Incas in the Valle Sagrado called their terraces *andinas* and were also constructed on the side of hills and mountains. Fields in the valleys were on flatlands and often irrigated with water from the Río Urubamba, known also as the Wilcamayu or Vilcanota, which means "sacred river." The terraces, however, were watered not from this river but from smaller ones that were high enough to flow directly into the canals.

The original Inca name (preserved in today's Quechuan) for the area was Tawantin Suyu, which means "the four regions." *Tawa* means "a group of four things," and *suyu* means "region" or "province"; these regions were Chinchasuyu, Antisuyu, Contisuyu, and Collasuyu. Their meeting place was in today's Cuzco or Cusco, called Quesqu, Qusqu, or Qosqu by the Incas.

The Incas set their most important cities and forts in this region due to the geographic and climatic advantages. Many cities were strategically located on high hills or mountains, where they were easily defensible. This protected Tahuantinsuyo, the Inca capital, from threats from the wild jungle tribes such as the Antis.

Choquequirao: The "Sacred Sister of Machu Picchu"

The "Sacred Sister of Machu Picchu," the city of Choquequirao is one of the least known of the large Inca ruins. It has been called a sister of Machu Picchu due to their many similarities, especially in architecture. It was first mentioned by Cosme Bueno in a Spanish document from 1768, then later reencountered by Eugène de Santiges in 1834 and finally mapped in 1837 by Léonce Agrand, France's consul in Lima. Then in 1909 American historian Hiram Bingham again came upon Choquequirao, located in Salcantay mountain range. Recently rediscovered, the site is

Arequipa, Peru. Photograph by Mylene d'Auriol Stoessel.

only partially excavated, and very few people ever visit it due to its remote location.

In the valley below the mountains runs the Apurímac River, similar to the Vilcanota running below Machu Picchu. Choquequirao is located at a height of 9,300 feet, which makes it less accessible than Machu Picchu. The site covers approximately 12,600 acres, and about 40 percent has so far been uncovered and recovered from the wild vegetation. It contains temples dedicated to the worship of the Sun God, Inti.

Scholars say that Choquequirao, which means "Cradle of Gold" in Quehuan, was ordered to be built by Túpac Inca Yupanqui, the son of Pachacuti, who built Machu Picchu. It was in the valley west of Vilcabamba, a gateway to Vilcabamba. Vilcabamba means "sacred valley"; the name is derived from two Quechan words, *huilco*, meaning "sacred" or "god," and *bamba*, meaning "valley." Some note that

Choquequirao was located at the tip of a triangular-shaped part of the Vilcabamba, which gave the Incas control over the entrance to the valley and prevented the Spanish from entering; they fought for decades before they were able to conquer the "sacred valley."

Some specialists even believe that Choquequirao could have played an important role in commerce and transportation between Cuzco and the Amazon region. It is speculated that Incas periodically traveled to the Amazon basin and brought merchandise back from there, even creating settlements there. Another theory holds that Choquequirao was a remote religious center of the Incas. Mountain gods and other deities could have been worshipped there.

Work to uncover the city started in the 1970s, but unfortunately many of the objects that previously belonged to it were already gone. One of the objects still intact is a big rock, with grooves carved into it that form a map of the irrigation system in this part of the Colca canyon. If water is poured onto the rock it runs through the grooves in the same way it runs through the irrigation system. The site still has the remains of Incan water wells built around 1600.

The Terraces of Moray: An Agricultural Laboratory

Since the Andes valleys are deep and narrow, there is little level ground suitable for irrigation at their bottoms. Those of us who have worked with terraces know that they are prone to landslides during rainy seasons, but the Incas solved the problem by terracing the whole valley. One of the added benefits is that terracing provided new land for growing crops and thus helped diversify their crops.

The reason for terracing, according to scholars, besides improving soil fertility, was to see how far they could expand the growing of corn and their irrigation techniques. Terrace building reached its apex before the coming of the Spanish, though it probably started in Lake Titicaca around 900 BC. Cocao, which is what produces cocaine and also Coca-Cola, according to historian Edward Ranney, was planted on special lots. Some terraces were also used for growing herbs, spices, and other medicinal plants. Terracing, which was used to grow corn, potatoes, and

Moray, Peru. Photograph by Mylene d'Auriol Stoessel.

tomatoes, gave the Incas the ability to feed an estimated population of ten million people.

These terraces, separated by *pirkas* or uncemented stone slab walls, resemble giant stairways. By filling them with fertile soil they not only took advantage of the rain but helped to prevent landslides, combining functionality and beauty.

At an altitude of 10,500 feet and forty miles from Cuzco is a great archaeological complex named Moray, composed of four circular terraces. According to scholars they were constructed by the Aymara-speaking culture from Tiahuauaco, Bolivia. Though no one knows what uses they had in mind for these circular terraces, experts are of the opinion they were intended to be used as an agricultural laboratory in order to experiment with crops at different altitudes, similar to the almunyahs introduced into the Iberian Peninsula by the Moors. The Moray circles are built in such a way that one can climb up and down the concentric levels using starlike stones that were implanted into the sides of the terrace walls. Due to the depths of the terraces, the lower levels have lower temperatures and therefore somehow simulate high-altitude conditions. Plants that prefer low temperatures could be grown in the lower terraces, where it is cool enough. The temperature difference between the lowest level and the highest one is said to be 59°F, the same as the temperature difference between sea level and an altitude of three thousand feet.

The Incas were very advanced in their use of farming and irrigation techniques, and they deliberately designed growing areas by microclimate so they could grow a variety of crops. This enabled the Incas to grow over 250 plant species, using complex irrigation systems, at different climatic conditions. (It's interesting to note that in New Mexico farmers are very aware of this concept of planting different types of plants at different altitudes. That's why they usually don't plant fruit trees in the low areas close to the river, for example, since they are prone to late spring freezes.)

Moray was first seen by non-Incas in 1932 when Shirppe Johnson's expedition flew above the site, twenty-one years after Hiram Bingham first saw Machu Picchu, though Moray is more accessible. Exactly why the

Incas put so much effort into building Moray is still not known, but there are many theories. So much earth had to be moved by supposedly primitive methods that some theorists have even asked if the terraces were a result of a meteor or a natural sinkhole. More than likely, the Incas took advantage of natural depressions in the terrain, which they then shaped based on an architectural plan. This included the construction of a network of aqueducts and drains for irrigation and release of water brought by the rains. It was so perfectly built that it still works today. One thing experts agree on is that Moray was an agricultural complex. Today on the Maras plateau, potatoes, lima beans, wheat, and barley are grown; little corn is planted due to the fact that the weather is too cold for this crop.

Though the derivation of the name Moray is not clear, some experts say it is linked to the harvest of corn or maíz (*aymoray*), and others say it refers to the dehydrated potato (*moraya* or *moray*). Still others claim that it's related to the month of May.

Throughout the years, many explorers have analyzed the site and come up with a variety of theories about its origins and purpose. Historian Victor Angles says that at some point the inhabitants of the plateau and the gorge became enemies, interrupting trade. For this or another reason, Angles says, the ancient settlers needed more corn and thus needed to prepare more land for planting. They decided to dig giant furrows to warm up the land in order to grow corn. This could have been the reason for the construction of this important complex. Until recently, the farmers of Moray grew corn, but it was later prohibited in order to protect the terraces.

John Earls, a researcher into Andean history, said he has come across "vertical stones in terraces" at Moray that would have served to mark the limits of shadows at dusk during the equinoxes and the solstices. Earls also affirms that the different terraces of Moray reproduce the different temperature areas of the Inca Empire. However, this seems evident and inevitable: the vast empire had different climate areas with specific temperatures. Because Moray is layered, sunken down into the ground, it is obvious that there will be lower and higher temperatures there. The Incas also cultivated wild plants at Moray, domesticating them and encouraging them to be more productive.

The light of the sun is said to be perfect on the sides of the mountains at Moray, and the agricultural terraces are solidly built. The Incas also paid careful attention to the different types of soil they used to fill the terraces, in order to irrigate them as well as possible and drain the surplus water. Experts claim that the various layers at Moray represent at least twenty different climate categories, making it suitable for the growing of almost any plant from the Inca Empire. Others say the terraces could have been used for estimating the production of certain plants. The site seems to be nothing more than a sophisticated, layered agricultural system with terraces and aqueducts that still function today. It is believed that the Incas would experiment, especially with the hybridization of vegetables, to see whether some plants were resistant to certain temperatures, or not.

Journalist Alfonsina Barrionuevo describes the monument in the following way:

A unos 7 km de Maras se encuentran los jardines colgantes de Moray construidos en un hoyo gigantesco de tierra. Una serie de andenes circulares descienden hasta una profundidad de 150 metros. Allí los Incas cultivaron maíz, quinua, panti, flores de kantu y otras plantas en vías de experimento para recreo de sus señores. Un sistema de canales donde hoy se deposita la lluvia aseguraba el regadío de las terrazas colgantes del gran anfiteatro. Moray fue sin duda un paraíso artificial de plantas y flores, algo más como un invernadero abrigado en el mismo interior de la tierra.

(Seven kilometers from Maras lie the hanging gardens of Moray, built in a gigantic trough. A series of circular platforms descend to a depth of 150 meters. There the Incas grew wheat, quinoa grain, panti, kantu flowers, and other plants as part of an experiment. A system of canals where the rain accumulates today ensured the irrigation of the hanging terraces of the amphitheater. Moray was no doubt an artificial paradise of plants and flowers, something like a greenhouse within the ground.)

Scholars say the Incas used the experience they obtained through their experiments with this type of agriculture, which can be said to be similar to a greenhouse, to organize the agricultural production of what they called the Tawantinsuyo, an experimental garden. In a way it can be said that Moray was similar to the almunyahs of the Arabs, experimental and recreational gardens used to acclimatize different vegetables and trees in the Iberian Peninsula, although the almunyahs did not necessarily use sunken, steeped terraces as in Moray.

The Tipón Aqueducts

Another agricultural research site was Tipón, located east of Cusco, which is considered an engineering masterpiece of planning, design, and construction, especially regarding its irrigation system of canals, aqueducts, fountains, buried conduits, channels, cascades, and artificial gullies. Tipón is an example of the great number of building and engineering feats of equal or even greater architectural and functional beauty that exist in the area. Questions remain, however: How did these ancient settlers manage to achieve this level of development? What drove their creative efforts? And how did they register their discoveries and progress? Someday archaeologists and other scholars might be able to answer these questions; for now visitors can only marvel at their beauty. The water channeled via these stone structures was utilized for irrigation, but also for baths and ponds. This indicates the Incas' knowledge of hydraulics and shows that, like the Romans, they were masters of water systems.

The *Camellones* or *Andenes*

Dr. Clark L. Erickson of the University of Pennsylvania, who has studied this pre-Columbian landscape, writes in the essay "The Lake Titicaca Basin: A Precolumbian Built Landscape," in *Imperfect Balance: Landscape Transformations in the Precolumbian Americas*, "The construction of raised fields (*waru waru, suka kollas*), stone-faced terraces (*andenes*), sunken gardens (*q'ochas*), irrigated pasture (*bofedales*), and a multitude of features related to the infrastructure of agriculture and settlement are essential elements of this anthropogenic landscape." In

Tipón Canals, Peru. Photograph by Mylene d'Auriol Stoessel.

Spanish these raised fields are known as camellones and in Aymara as suka kollas. Erickson says that he believes the cultural landscape was constructed piecemeal by Aymara and Quechua farmers who practiced small-scale intensive agriculture. This human-built landscape, similar to the man-made acequias, is located in the Andean cordilleras and over time has been transformed into an irrigated agricultural landscape.

The basin possesses three main landscape features: terraces, raised beds, and sunken gardens. It is said that the most striking feature is the patterned landscape of terraces—*terrazas* or *andenes* in Spanish, *pata* in Quechua, and *takha* or *takhana* in Aymara.

The raised fields, or camellones, are large elevated planting platforms built on areas of waterlogged soils, similar to ciénagas, or areas prone to annual flooding. They mimic the chinampas of Xochimilco, or Les Hortillonnages in France and Ses Feixes on the island of Ibiza, in that the platforms have canals on one, two, or all sides that were created during the process of raising the field. These raised fields are normally twelve to thirty feet wide and thirty to three hundred feet in length. The canals improve the microclimate in that the water in the canals functions as a solar heat sink. This reduces the risk of frost damage by increasing the temperature.

Sunken gardens—or *q'ochas*, Quechua for "container of water"; *q'otanas*, *cotaña*, or *cota* in Aymara; and *charcas hundidas*, *pozos*, or *ojos de agua* in Spanish—are the third major element of the landscape. Q'ochas are large shallow depressions usually located in poor drainage areas. The forms, covering up to ten acres and up to eighteen feet in depth, are round or oval shaped, and they are ranked according to size and shape. The large and round form is the most common and is known as a *muyu q'ocha*, while the medium-sized and oval is a *suyt'u q'ocha* and the small rectangular one is a *chunta q'ocha*.

Another type of landscape in the basin is the irrigated pasture. Llamas can survive on the high-altitude grassland, or *puna*, in the basin, but alpacas require a more succulent forage than the dry, tough grasses of the puna. Pasture for alpacas can be found only in the Distichia moors or *bofedales*, *oqho* in Aymara. The basin's bofedales are fragile and require regular maintenance. Water is brought from rivers, streams, and springs in two large feeder canals—*hach'a irpa*. These in turn supply smaller networks of canals—*hisk'a irpa*—which are used to create the bofedales. The system is similar to that of the *acequia madre* and the *acequia secundaria*.

According to Erickson,

> Many of the Precolumbian technologies, landscapes, and indigenous knowledge systems [in the basin] are abandoned, underutilized, or forgotten. Their physical imprint on the landscape and built environment is enduring. Despite dramatic changes in land tenure, demography, social organization, and economic systems during the past 500 years, the Precolumbian structures of everyday life (fields, pathways, walls, canals, and other features of the built environment and landscape) still shape contemporary rural life in the region. This valuable record of Andean environmental history is now at risk as terrace walls are removed for pasture and raised fields are erased by mechanized plowing and urban expansion.

The Acequias of Argentina

Acequias at one time seemed to line every street in Mendoza, Argentina, allowing for tree-studded avenues to stretch throughout the city, which is situated in a desert on the eastern side of the Andes, at an elevation of 2,500 feet. The acequia system was built on foundations laid by the local Huarpes tribe. The acequias make Mendoza the wine capital of South America, rivaling both France and California. The red wines are best. Malbec is particularly flavorful.

When settlers under the Spanish Crown first arrived in the inland region that is now called Mendoza, they found a very green and fertile land in the midst of a vast desert. The indigenous Huarpes had, many years earlier, developed and built a complex and sophisticated system of irrigation channels to bring water from the Mendoza River to the arid plains. These channels, which later became the acequias of Mendoza, featured advanced hydrodynamic techniques—which suggests that the Incas might have controlled the area—that regulated the flow of water, allowing for efficient use of the scarce resource. The Río Mendoza depends almost solely on snowmelt during spring and summer. Today the irrigation system inherited from the Huarpes supplies both Mendoza's residents and all its viticulture lands with water. Irrigation channels have been extended and added to, but the basic structure of the system remains the same. Water is rationed between vineyards and farmers through the opening and closing of miniature flood-control gates. On the vineyards growers employ similar techniques to flood irrigate around the base of their vines. This simple flood-irrigation technique has been used for centuries in Mendoza and is only now, in a very few cases, beginning to give way to more complicated and expensive drip-irrigation systems.

Dr. Jorge Ricardo Ponte, a sociologist by training, has written two books on the acequias of Mendoza, which he refers to as the "culture of oasis." I had the opportunity to meet Dr. Ponte at a conference in Mexico City in 2006, and he visited Embudo in 2010 to study the acequias. Dr. Ponte cautions that we know little about hydrology in pre-Hispanic times, as Native groups left no written documentation. Scholars rely on oral histories that were recorded between 1575 and 1696. But he goes on to say that there were acequias in Río de Cuyo prior to the arrival of the Spanish. These acequias were born on the Toma del Inca (known to the Huarpes as Goazap-Mayu) and were named the Acequia de Tabal-que, the high Acequia de Tantayquen, the Acequia de Allayme, and the Guaimaien *sequia*. Documents also mention other acequias outside of the new city. There are also acequias in Tilcara, Argentina, on the margins of the Río Grande, according to Miguel Angel Martiarena, who holds a master's degree in landscape architecture from the Catholic University of Cordova.

Mendoza was known as the Valle de Huentata, and it was the southern frontier of the Inca Empire. Ponte writes that in the "imaginario social mendocino" it is believed that the Incas helped the Huarpes to reorganize their hydrology system. This is repeated in local histories, though there are no documents that will prove it. But since the Incas are recognized as having superior technical abilities to the Huarpes it can be supposed that they indeed helped after their arrival around

1481. Though the Incas were used to planting in terraces, the lands in Mendoza are not so steep and mountainous and the Huarpes were used to doing flood irrigation, or *por mantos*. This was the optimal system of irrigation for the cultivation of corn, squash, *porotos, zapallos*, potatoes, and other vegetables that were part of their diet. In the actual city of Mendoza, according to a 1566 document, there were four acequias, la de Allayme, la Tabal, la de Gaimaien, and "la que pasa por este pucará" ("fortress" in Quechua).

On his website, Dr. Ponte writes, "The Mendoza of the acequias, the result of a particular and exclusive historic development, has become in the course of time a very exciting urban and hydrology model. From there, the width of its streets, the existence of *'acequias callejeras'* that border its trails, the existence of trees full of leaves that flank its avenues, have not only served for the orientation of the growth of the city over its agricultural suburb but [have] become the model for the rest of the cities in the province of which Mendoza is the capital."

After an earthquake destroyed Mendoza's sister city of San Juan— founded in 1562, one year after Mendoza, and until 1820 part of the province of Cuyo, and also a *ciudad de oasis*—in 1944, the city was reconstructed using modern Mendoza as the model. "Therefore, at present, in the City of San Juan, wide streets and trails are the norm, with trees along its avenues, with urban acequias part of the urban layout. This wasn't part of the historic San Juan before the destructive earthquake," Dr. Ponte explains. "The Mendocino hydrologic system of canals and acequias not only constitutes the productive support system but it also has to be seen as a cultural patrimony, having in that category all of the necessary requirements of the cultural benefits of a community."

MESOAMERICA

The *Chinampas* of Xochimilco

One has to travel over several thousand years and across continents, from the first spring-fed irrigated gardens at Jericho to the chinampas

Chinampas, Xochimilco. Photograph by the author.

of the Aztecs in Xochimilco, in order to put in perspective the agricultural revolution that occurred starting in 1492. We cannot attempt to comprehend traditional agriculture without exploring the Mesoamerican connection. In discussing Indo-hispano agricultural traditions, many times scholars, especially those who continue promulgating La Leyenda Negra, or the Black Legend (a term coined by the Spaniard Julián Juderías to describe the depiction of Spain and Spaniards as "cruel," "intolerant," and "fanatical" in anti-Spanish literature, starting in the sixteenth century), simplistically and erroneously continue referring to "Spanish" origins, encompassing Castilian, Christian, European, and other traditions they don't fully understand.

A detailed plan of Mexico City by Juan Gomez de Trasmonte dated 1628 shows the *zócalo*—the town square—and the acequia, which ran from east to west along the south side of the Royal Palace, today the National Palace. The evidence regarding the Acequia Real, which went from the Calle Corregidora to the Alhóndiga and Roldán, demonstrates the importance of archaeology for recuperating historic spaces that are under the actual city. Archaeological study also helps to reevaluate the acequia's daily use in different historical epochs. Recently the destruction of the colonial acequia located in the Calle Corregidora has been covered in the media. This has to do only with the re-creation of some old bridges and canals used for navigation purposes in Mexico City. During excavations completed between October 1980 and April 1981, part of the acequia was excavated extensively from the start of the Calle Corregidora—previously Meleros, Puente de la Leña, Pulquería de Palacio, and de la Acequia—all the way to the National Palace, where the acequia meets Pino Suarez at its crossing with Calle Roldán and the Calle Alhóndiga.

I've had the privilege of visiting the chinampas of Xochimilco on several occasions, and exploring, not where the tourists go, but deep inside to see the last few agricultural chinampas that still produce food for the market. The word *Xochimilco* comes from the Nahuatl *xóchitl*, which means "flower"; *milli*, a "cultivated place"; and *co*, "place," or a place for cultivated flowers. And *chinampa* comes from the Nahuatl *chinamitl*, which means "hedge or enclosure of canes" or "reeds", and *pan*, "over the," because in the beginning the chinampas were made of reeds, then filled with soil from the bottom of the lake. This gave them the appearance of being floating gardens. The Spaniards called them *sementeras* or camellones; they are also known as *tajones*. The chinampas landscape, same as that of the acequias, is a testimony to the evolution of social creativity, the imaginative and spiritual vitality of the people who have labored to create such bountiful landscapes. Angel Palerm, one of Mexico's most renowned scholars, says there are two kinds of chinampas, those he calls the "lacustrine chinampas," which were artificial islands built in the lagoons, and the "dry chinampas," which were located in marshy zones where drainage was a problem. These kinds of chinampas seem to have existed in the environs of Cholula and the state of Puebla.

Dr. Saúl Alcántara Onofre, landscape architect and professor at the Universidad Autonoma Metropolitana in Mexico City, says, "A chinampa is, generally speaking, a floating or fixed garden formed artificially either by dumping earth in a designated area near the shore of a lake until a small islet is formed, or by making a sort of raft of logs, rushes and similar materials, on which earth and compost is then laid, until the raft gradually sinks and touches bottom." He adds, "A basic characteristic of the chinampa cultivation system is that canals located between the artificial islets not only serve as circulation passages but also for the provision of water. This arrangement results in an extraordinarily fertile and highly productive agricultural pattern."

As already noted, this type of agriculture has appeared in different parts of the globe, under different names, at different times in history. In each place it developed in marshy areas that had to be drained in order for them to produce food. Some are at the edge of the sea, others on the edge of a river, others in lakes that were drained, and yet others in high-altitude marshlands. Ses Feixes of Ibiza developed at sea level, Les Hortillonnages of Amiens, France, at around 100 feet above sea level, the chinampas in Xochimilco at around 7,300 feet, and the camellones in the Andes at over 12,000 feet in elevation. Though UNESCO protects three of them as world heritage sites, all four are but a remnant of what they were, and though they are still used for some type of agriculture, they have all become tourist destinations, with the exception of the camellones.

Partidores, Aldama, Chihuahua, Mexico. Photograph by the author.

PART 2

THE
KNOWLEDGE
OF THE
WATER

El agua es un don divino, porque sin agua no hay vida.

2

The Camino Real de Tierra Adentro

The Water Road

≈

NOW THAT WE HAVE taken a tour of other community irrigation and agricultural systems in different parts of the globe, we are ready to embark on the next part of our journey. What came to be known as the Camino Real de Tierra Adentro (the Royal Road of the Inner Province) has been used by the people of the Americas since prehistoric times. During the Spanish epoch, the road stretched from Mexico City to Taos, though most historical accounts have it ending in Santa Fe. The only problem is that Santa Fe didn't exist when Oñate made his way up north, since he settled in San Gabriel, on Ohkay Owingeh land, about twenty-five miles north of present-day Santa Fe. Until the Santa Fe Trail came into existence all trade and migration into northern New Mexico was from south to north. But the Camino Real has also been called the Camino de Agua, that is, the Water Road, because the settlers, especially those from Zacatecas on north starting in 1596, had to make sure as they traveled that there was water for their animals and also for them.

El Camino de Agua, the Water Road, can be traced by the names of settlements such as San Juan del Río, Aguas Calientes, Ojo Caliente, and Ojo de Talamantes and also by the names of places where the water became scarce, such as the Jornada del Muerto, the "Journey of the Dead," in what is now southern New Mexico. But as they traveled north they were also very much aware of where there were chupaderos, pozos, *manantiales*, arroyos, tinajas, and *socavones*; whatever could hold water or any place where there was water that could be had, they knew about it. When we stopped at Valle Allende on our way to Durango for the 1997 Coloquio del Camino Real de Tierra Adentro, Rita Soto, a local historian from Valle de Allende, explained to me that settlers had to have come through "El Valle" instead of Santa Bárbara for the simple fact that there was more water in San Bartolome, as the place was initially known, because of its proximity to the Ojo de Talamante, which is fed by forty manantiales. Though she didn't refer to the Camino Real as the Camino de Agua, she did stress that "the settlers had to follow the water sources in order to survive." The Ojo de Talamante is a beautiful spring close to El Valle de Allende that I had the privilege of visiting for the first time in 1998. Now it has become a swimming hole and a place for picnics for the local people. Today the Ojo is starting to dry up due to urban pressure, as wells are dug to take water to the city of Parral, northwest of El Valle de Allende. Parral, a city with a population of about one hundred thousand, is the sister city of Santa Fe, New Mexico, and the place where Mexican Revolutionary hero Pancho Villa was assassinated on July 20, 1923. Walking the streets of Parral one can almost sense that Pancho Villa is still alive, as everywhere one goes, whether to a local cantina or simply to the streets, people still talk of him as if he had just walked by.

But no one has stressed more that the Camino Real is actually the Camino de Agua than Dr. Tomás Martinez Saldaña, who has devoted his life to studying the role of the Tlaxcaltecas in the development of agriculture in northern Mexico, including present-day New Mexico. In a recent personal communication he wrote, "El camino real es una especie de escalera hidraulica o acuatica, donde cada parada, cada lugar, cada espacio tiene un referente al agua" (The Camino Real is in a sense a hydrologic or aquatic ladder, where each stop, each place, each space is somehow related to water).

The old Camino Real between Ohkay Owingeh and La Joya, today Velarde, followed what today is known as the "camino del medio," which goes through Alcalde (formerly La Soledad del Río Arriba), Los

Medida Antigua, una naranja de agua, Valle Allende, Chihuahua.
Photograph by the author.

Pachecos, La Villita, Los Luceros, and Las Cachanillas. It then made its way to the Plaza del Embudo (Dixon today) following the south side of La Mesita instead of traveling around it along the Río Grande, then down the Cañada del Embudo, today known as Arroyo de la Mina, through the plaza, then up to the Apodaca Trail, where it forked, with one road going to Picurís and the other to Taos. This part of the road (from Ohkay Owingeh to Taos and Picurís) has become lost in the effort by historians on both sides to tie the road from capital to capital, Mexico City to Santa Fe.

THE HISTORICAL DEVELOPMENT
OF ACEQUIAS IN NEW MEXICO

Now we will zero in on the acequias of the American Southwest, especially those in northern New Mexico. Prior to the arrival of colonists under the Spanish flag, the Native Americans of the Southwest had been using irrigation for centuries, including one ancient canal utilized in the modern settlement of Mesa, Arizona, in 1878, according to Wells A. Hutchins in his January 1928 article in *Southwestern Historical Quarterly*, "The Community Acequia: Its Origins and Development." The explorers Francisco Coronado and Antonio de Espejo spoke of irrigation canals in the Middle Río Grande Valley. Though not much is known about Native American irrigation institutions, given that the pueblos are by nature communal, taking care of their canals must have been a community effort. Regarding the communal systems of the Native Americans, Law 4, Title I, Book 2 of the Laws of Indies provided that the old laws and customs that the Native Americans enjoyed should be retained and respected so far as practicable. As a result, since both systems were in a way compatible, in some instances the New Mexico acequias show distinct traces of the influence of the Native American customs.

When the colonists under don Juan de Oñate arrived in today's Española Valley and settled on Ohkay Owingeh land in San Gabriel, today Chamita, in 1598, the irrigation knowledge they brought with them had been acquired from the Moors, who in turn had gotten that knowledge, it appears, from Yemen and the Indus Valley. But the Native Americans and the Moors had one thing in common: the water was seen as a communal resource and not as a commodity.

POLICIES THAT INFLUENCED THE SPANISH COLONIZATION PROCESS

Now let's take a closer look at how the acequias in what today is the southwestern United States and northern Mexico developed and how they ended up taking different paths after 1848. Four major documents influenced the development of water-related policies during the Spanish colonization process. The Siete Partidas, promulgated by King Alfonso the Wise in the thirteenth century, to this day still influence water law in New Mexico and even in places such as Louisiana. The first mention of easements is found here. The Third Partida, Laws 3

Acequia, Valle Allende, Chihuahua. Photograph by the author.

and 6, Title 28, declares rainwater as common, as well as the use of water in the rivers. And Title 31 deals with *servidumbre*, or easements, as they relate to the acequias and aqueducts. It says that "such easement shall be twice as wide as the measurement of the *alcabús*, or the inside of the canal, or *cuatro pasos de Salomón* (four steps), measured on each side of the acequia, of which easement, no person can claim it as private property since it is community property." A "paso de Salomón" is defined as one and two-thirds *vara*, or 55 inches, 4.58 feet. A vara is approximately 33 inches (definitions vary: a vara is 32.992 inches in Mexico, 33 in California and New Mexico, 33.372 in Florida, 32.993 in Colorado, and 33.33 in Texas). Therefore the easements on each side of the acequia would be approximately 18 feet on each side for an acequia that measured 9 feet wide on the inside. Most of the acequias in New Mexico are half as wide, so the banks on either side should be around 9 feet. It must also be noted that in the past, before the road system was developed, the banks of the acequias were used as roads. Another important thing to note about the Partidas is that Law 4, in close approximation to Roman law, states that floodwater should be directed away from acequias either by an open channel or underground, so as to prevent damage to neighbors. These two issues, the easements and what to do with arroyos, cause more challenges than anything else in acequia communities today. First, a lot of people, especially newcomers who move into the acequia communities, don't want to honor the easements. Second, no one wants arroyos through their properties, and as a result most arroyos empty directly into the acequias, causing them to silt up, and then the cost of repair has to be borne by all the property owners.

Another important text was the "Ordenanzas de descubrimientos, nueva población, y pacificación de las Indias" of July 13, 1573, which is the document don Juan de Oñate had to abide by when he signed his contract in 1596 to colonize New Mexico. The best-known of these documents, based largely on the Partidas and Ordenanzas, was the *Recopilación de las leyes de los reinos de Indias* of 1681 (the Laws of the Indies). Another influential document was the Plan of Pitic of 1783, based largely on Laws of the Indies; this was supposed to be the model document for all settlements on the northern Mexican frontier after that date. At that time New Mexico was part of Nueva España and definitely part of that northern frontier. In fact, Taos, New Mexico, was settled following the Plan of Pitic.

Writes Hutchins,

> Thus we find that the Partidas allow a town resident to construct a private diversion acequia provided it does not interfere with the common use of the town. The Laws of the Indies provide that all waters in the Indies shall be common to all inhabitants (lib. iv, tit. xvii, ley 5); that the viceroys shall inform themselves concerning irrigable lands and cause them to be sown to wheat and not grazed by cattle (lib. iv, tit. xii, ley 13); that the distribution of lands and waters to settlers shall be made on the advice of the village councils (lib. iv, tit. xii, ley 5); that the Indian rules governing water distribution shall be maintained among the Spaniards to whom the lands have been assigned, each to be given the water in turn (lib. iv, tit. xvii, ley 11); distinguish between irrigable and non-irrigable lands in the laws on colonization (lib. iv, tit. vii, ley 14); and direct the viceroys and the courts to make provisions regarding the waters and other public things in order best to promote the public welfare (lib. iv, tit. xvii, ley 9). The *Novisima Recopilación* of Spain and later decrees relating to the Indies directed and encouraged public irrigation works.

THE *REPARTIMIENTO,* OR SHARING OF WATER

The repartimiento, or sharing of the water, is a Moorish concept that dates back over four thousand years, with its origins in Assyria. The concept was also used by the Romans on the Iberian Peninsula between the fourth century BC and the fifth century AD when there was a shortage of water, according to *Sharing Water: Irrigation and Water Management in the Hindukush—Karakoram—Himalaya,* edited by Hermann Kreutzman. In the repartimiento we can find the genesis of water sharing, the same practices found throughout the world in arid lands.

The repartimiento is done according to both volume and time. By volume the water is divided into *bueyes*, *surcos*, and *naranjas*, and by time it is divided according to hours, or even quarter hours. For example, in the Embudo Valley eight of the ten historic acequias that have senior water rights share water in times of shortage, or *escasez*. Usually the upper four acequias have the water for three days, then the lower four for four days. The water should be shared based on the number of acres each acequia has under irrigation. Within each individual acequia, the water is again shared by all the *parciantes* (water-rights owners), based on the number of acres each parciante has rights for. Sometimes conflicts arise, however, because the small acequia parciantes end up getting more time to irrigate than the big acequia parciantes. If a parciante has one *peón*, or one full right, then that person should be able to irrigate more than someone who has only one *medio-peón*, or half a right. In practice it doesn't always work this way. In an acequia with ten parciantes and twenty-five acres, for example, each parciante will get more time per acre than a parciante in an acequia with two hundred acres and one hundred parciantes.

Sharing of the water is based on the old methods of measurement. A buey de agua is the amount of water that flows between the front and back legs of an ox up to its belly when it is standing in the middle of the stream. In reality it's a square vara, or 725 liters per second, or 191.58 quarts. A vara castellana is 32.8748 (linear) inches. And there are 48 surcos in a buey de agua; in New Mexico a surco was the amount of water that could flow through the opening of *a buje de una rueda de carreta*, or the center of the hub of a wooden cartwheel. It's roughly the amount of water that fills a five-inch-square area. A surco, then, is 15.1 liters or 3.99 gallons per second. A surco is also defined as the amount of water that is needed to irrigate a suerte of land, approximately thirteen acres, in a twenty-four-hour period. A suerte of land is two hundred varas in width by four hundred varas in length.

Possibly the most ancient of practices regarding acequias, whether in the Middle East, the Iberian Peninsula, Mexico, or New Mexico is the ancient ritual of cleaning the canal every spring, known as the *limpia* or *saca*. Also, the role of the mayordomo has not changed much, and the water is still used today to grow food just as when the first acequia, wherever it might have been, was constructed. Easements, which today cause so much consternation especially among the new people moving into acequia lands, have been recognized since the time of King Alfonso the Wise of Spain in the mid-1200s. The anatomy of the acequia is still basically the same today whether in the southwestern United States or in Murcia, Spain. Water is diverted by the use of an *azzud*, in New Mexico known as a *presa*, or dam, and channeled along the acequia madre (mother ditch) and finally put back onto the river by the use of a *desagüe* or *escurrentina*.

WATER RIGHTS AND THE TREATY OF GUADALUPE HIDALGO

There is probably no more misunderstood and misinterpreted section of the New Mexico State Constitution than Article II (rights under Treaty of Guadalupe Hidalgo preserved) when it comes to water rights. Hundreds of thousands, if not millions, of dollars have been spent paying scholars and lawyers to clarify what rights we have under the treaty. Any attempt to understand what those rights are regarding the use of water for irrigation come from the Native Americans, Romans, Visigoths, and Arabs.

According to Wells A. Hutchins,

> Many acequias had been built in the territory acquired by this nation from Mexico. The Treaty of Guadalupe Hidalgo and the Gadsden Purchase Treaty provided that Mexicans living in the ceded territory should be protected in their enjoyment of liberty and property. So far as community acequias are concerned, what these treaties did was to protect effectively the valid water rights of acequias then existing. They did not, however, continue the Mexican laws in force; therefore, the continuance of any specific Mexican law or custom was a matter to be determined by proper legislative authority for each political subdivision carved out of the former Mexican territory.

Sacando Acequia Junta y Cienaga, Embudo.
Photograph by Donatella Davanzo.

In other words, the laws pertaining to the acequias in New Mexico are different than those in California, Arizona, Texas, or Colorado, for example.

Even before the signing of the Treaty of Guadalupe Hidalgo, the Kearny Code of September 22, 1846, upon the organization of a provisional government in New Mexico, provided that the laws concerning watercourses should continue. What changed was that the regulation of the water was transferred from the village *ayuntamientos* to the alcaldes and prefects of the different counties.

The first reference to, or attempt at, any form of water law is found in the Fuero Juzgo of AD 654, adopted by the Visigoths, which deals more with penalties for the abuse of water. "El Fuero Viejo de Castilla," part of the Fuero Juzgo, refers only to the use of running water for the functioning of gristmills and for fishing purposes. It is not until the thirteenth century that we encounter a series of laws dealing with the use and distribution of water for irrigation purposes, based on the ancient Roman law. These water laws were part of the Siete Partidas, a monumental effort to codify all existing ordinances up to 1256 that was the work of King Alfonso X, known as "the Wise." The publication of these laws signaled a step in the right direction in the cultural evolution of Europe and Spain. Though influenced by the Romans, they weren't put into practice until the Ordenamiento de Alcalá was enacted in the middle of the fourteenth century. The Tercera Partida declares "common things" to include "the flood waters and the use of the rivers." It also stipulates that "the headwaters that are found there" are common property.

Therefore, the Treaty of Guadalupe Hidalgo also protected the right-of-way of easements for the acequias, as defined by the Partidas. This is very important, for it means that the treaty defines legally for the first time the rights of acequias in relation to the private property through which they zigzag. Today many property owners try to block the mayordomo or peones from going through their property during spring cleaning of the acequias.

In New Spain, including New Mexico, the new legislation pertaining to land and water that started to emerge follows the Siete Partidas and is compiled in the Laws of the Indies. These laws stipulate, for example, that "the pasture, mountains and water shall all be communal." Others deal with administrative mandates, like the naming of water judges (today called mayordomos) to distribute the waters used for irrigation by the Indians. Yet others give the viceroys and courts the right to administer the waters in terms of "justice and equality." This is where the custom of sharing the water in times of shortage comes from.

On paper, at least, the interest and respect the Crown had for the "Indians" is evident in several laws. The Laws of the Indies order respect for the water rights the Indians had and decree that the waters should be shared equally among the Indians and Spanish settlers. For New Spain (which included modern New Mexico), the court in charge of administering the law was in Guadalajara. In 1788, in the ninth edition of the Laws of the Indies, King Charles III included language concerning the construction of new acequias where needed for irrigation purposes. A royal order by Charles IV, dated November 18, 1803, and confirmed four years later, protects communal rights over individual rights as established by the Siete Partidas and the Laws of the Indies, not only in general terms but in specific cases, holding "that the settler of such city is the true and only owner of the waters that run through public pipes, as long as the public needs them." This is also repeated in the Plan of Pitic of 1783. The Siete Partidas, the Laws of the Indies, and the Plan of Pitic are three very important documents that people need to understand if they want to know what rights were guaranteed by the Treaty of Guadalupe Hidalgo.

In 1851 and 1852, shortly after New Mexico became a territory, the legislature provided that courses of the acequias that were already established should not be disturbed and that the water should remain in common usage. The legislature also established that commissioners and a mayordomo should be elected by property owners and that the work of maintaining the acequias should be done in proportion to the amount of land owned, whether it was under cultivation or not. In 1880 the legislature added the provision that commissioners should measure the land served by their acequia in order to know how much labor each property owner had to provide. Further changes in 1895, 1897, and 1903 increased the powers of the commissioners and reduced the power of the mayordomo. The new legislation "allowed the commissioners to enter into contracts, provided that water should be withheld from delinquent landowners, declared all community ditches or acequias to be bodies corporate with power to sue or to be sued, and defined such community ditches as ditches not private or incorporated under State laws, but which are held by more than two owners as tenants in common or joint tenants," writes Hutchins.

In Colorado the territorial legislature passed a special law in 1866 "to regulate ditches used for farming purposes in the counties of Costilla and Conejos," and in 1872 it applied the same law to Las Animas County. This had to do with enforcing collection of assessments for acequias' upkeep. Meanwhile, in California, which was admitted to the union in 1850 without having been under a territorial government, the common law of England was adopted and Mexican law was excluded. According to Hutchins, "The legislature [of California] has never provided for the government of community acequias, except in so far as the acts incorporating the cities of Los Angeles and San Jose authorized the city councils to provide for irrigation." Though the southwestern United States at one time was part of Spain and then Mexico, under U.S. law, water law developed separately in individual states, and it seems that the New Mexico legislature, both territorial and state, retained many of the law's attributes from its colonial past.

We know that the first acequia in what is now New Mexico, using the definition of a canal based on Moorish concepts, was in the community of Chamita, at the junction of the Río Chama and the Río Grande, across from present-day Ohkay Owingeh. On August 11, 1598, a month after Spanish settlers' arrival and settlement in San Gabriel, work on the acequia was begun using the labor of 1,500 Native Americans. It is not known whether the Native Americans were all from Ohkay Owingeh or whether Tlaxcaltecas were also involved, since four hundred Mexican families came with Oñate.

Based on Hutchins's research, it appears that from 1598 to the Pueblo Revolt of 1680 a total of about 23 acequias might have been constructed in New Mexico. From the time don Diego de Vargas entered New Mexico again in 1692 to the end of the Mexican period upon the signing of the Treaty of Guadalupe Hidalgo, another 220 acequias were built, and from that period to 1909 another 247 were dug.

What is unknown, and not even lawyers seem to be able to answer the question, is, does the treaty apply only to those acequias in existence prior to its signing on February 2, 1848? Or are all acequias that are within part of a Spanish or Mexican land grant that has been accepted by the U.S. government governed by the treaty provisions, even if they were not constructed until after the treaty was signed? Another question is, what if a land grant was rejected? Are the treaty protections for the acequias still valid?

THE AGRICULTURAL REVOLUTION
OF THE RÍO ARRIBA BIOREGION

A similar agricultural revolution happened in northern New Mexico after don Juan de Oñate's arrival on July 11, 1598, as happened in 1492 when Columbus arrived on this side of the Atlantic or when the Moors crossed the Straits of Gibraltar and settled on the Iberian Peninsula in 711. To fully comprehend "traditional" agriculture in northern New Mexico, especially among the Indo-hispanos, these two other agricultural revolutions have to be understood, or at least looked at.

In the Río Arriba bioregion of northern New Mexico and southern Colorado, which extends from La Bajada south of Santa Fe north to the San Luis Valley, up to the mid-1930s (actually up to the advent of Los Alamos in 1943) most food was grown locally. My mother used to say that when she was growing up in the early part of the twentieth century the only items people bought in the local *tienditas* were *fósforos, aceite de lámpara, y azucar* (matches, kerosene, and white sugar). Now very little of our food supply is grown in the Río Arriba bioregion. If a food-shed is overlaid with a watershed, especially a microbasin like the

Embudo watershed—formed by the Pueblo, Santa Bárbara, Chiquito, and Trampas Rivers and the creek Ojo Sarco—it immediately becomes evident how little food is produced for local consumption. Chile, which has been grown here since it was introduced to the area in the 1580s, when the first seeds were brought over by Obregon during the Espejo expedition, more so than corn, is imported from the Mesilla Valley in southern New Mexico, and now northern Mexico, though at one time this was all part of Nueva España. Now, with Walmart Supercenters everywhere, globalization has arrived in what was once the outpost of Spain's northern frontier. Less than half of the irrigated land in the three-hundred-square-mile Embudo watershed is used for growing food; up to the Tewa Basin Study of 1935 close to 100 percent of the irrigated land was in agricultural production.

This region is the core of the Indo-hispano heartland, first settled by the Pueblo Indians and more recently by Spanish colonizers headed by don Juan de Oñate in 1598. Along with him came a contingent of Indians from the interior of Mexico. The Mexican scholar Dr. Tomás Martinez Saldaña, in conversations over the past several years and in his written work, argues that the Tlaxcalas probably influenced the agricultural development of the area more than any others. But as we'll see, the roots of our *agricultura mixta tradicional mestiza* are anchored globally.

THE RÍO GRANDE VALLEY: PUEBLOS ENCOUNTER NEW VECINOS

The Embudo watershed, which for the purposes of this book extends all the way to present-day Española on the west side of the Sangre de Cristo Mountains, was so named during the 1880s by the railroad industry as it promoted tourism to the southwest. But the watershed also spills to the eastern side of the southern Rockies because of a deal that the Picurís Indians made with the people from the new settlement of Mora in the early 1830s, allowing the settlers to take water from the Río Grande watershed and transfer it to the Canadian watershed. Currently nobody knows how the courts will rule when the water rights for the

Río Embudo and its tributaries are adjudicated by the Office of the State Engineer, nor do we know what stance leaders in Picurís and Ohkay Owingeh will take.

This watershed is unique, if for no other reason than because four major Spanish land grants at one time were controlled by the powerful Martín Serrano clan, who came to the area with Oñate in 1598. The Martín Serranos were mestizos. During the Pueblo Revolt of 1680 they ended up in El Paso del Norte, present-day Ciudad Juarez, but they returned with don Diego de Vargas starting in 1692. The most famous of the clan was Sebastián Martín, who was awarded what's known as the Sebastián Martín land grant in 1703, which was then reissued in 1712 and extends from present-day Alcalde (La Soledad del Río Arriba) east to Ojo Sarco and Trampas. Then, in 1725, when the Embudo grant was made, Sebastián's brother Francisco "El Ciego" Martín was one of three who applied and received the twenty-five-thousand-acre triangular grant, whose southern boundary is the Sebastián Martín grant. In 1751, when the Trampas grant was made, Sebastián allowed new settlers from the Barrio de Analco in Santa Fe to carve out part of his grant for the new settlement. Genealogy tells us that there were some Martín women intermarried within the clan. Finally, in 1796, when the Santa Bárbara grant was made, it was awarded to Valentín Martín, grandson of Francisco Martín, among others. Therefore, over a period of one century this family somehow came to control all the land between Ohkay Owingeh and Picurís. Again, based on documents uncovered by scholars, the Martín Serranos never denied their "Indian" blood. Today the surname Martinez is still the predominant one in the watershed and in the Española Valley.

Ohkay Owingeh was christened San Juan de los Caballeros by Oñate when he first came to the Española Valley in 1598. In 2006 the Indian village at the confluence of the Río Chama and the Río Grande reverted to its original name of Ohkay Owingeh. The village is located in north central New Mexico in the center of an area known as the Tewa Basin. It is situated on an eroded alluvial remnant about one mile east of the Río Grande and has been continually occupied since about AD 1300. At present the reservation covers about 12,213 acres, including about 1,800 acres of irrigated farmland. It is situated twenty-five miles north of Santa Fe and forty-three miles south of Taos.

Prehistoric plant remains reveal that the region's Native inhabitants grew a short cob of ten- to twelve-row corn, common beans, bottle gourds, two species of squash, and cotton. They also gathered piñon nuts, prickly-pear cactus, yucca fruits, juniper berries, pigweed, goose-foot seeds, and purslane. Purslane is a green that appears to have been native to both sides of the Atlantic, as it is also consumed in the Mediterranean area.

Ohkay Owingeh is located on the Upper Sonoran life zone. To the east are the Sangre de Cristo Mountains, while the Jemez Mountains are situated to the west. Elevation is 5,660 ft. The historic village is built of adobe and forms two plazas. Besides their house in the village, families also maintain a summer home in the agricultural fields. The language spoken is Tewa, and Ohkay Owingeh is considered the mother village. Tewa is a subfamily of Tanoan, a language family in the Uto-Aztecan stock.

Representatives of two of the first three major expeditions under the Spanish Crown reached Ohkay Owingeh: Captain Francisco de Barrionuevo, scouting for Coronado in 1541; and, in 1581–1582, the ex-pedition led by Francisco Sanchez, called "El Chamuscado," and Friar Augustin Rodriguez, the first Spaniards known to have visited the Pueblo Indians since Francisco Vásquez de Coronado forty years ear-lier. Only Espejo in 1582 did not visit Ohkay Owingeh. The first suc-cessful colonization in the Río Grande area was under Oñate at Yungue Oweenge, first renamed San Francisco and a year later renamed San Gabriel.

JUAN DE OÑATE AND THE ESTABLISHMENT OF SAN GABRIEL

With the arrival of Oñate and his colonists, which included 129 families who were either peninsulares (gachupines), criollos, or mestizos, there also came four hundred Mexican Indian families who were under contract with the Spanish Crown. The Tlaxcaltecas, who were never conquered by the Crown, were contracted to do the layout of the acequias and develop the agriculture of the area, according to Mexican agricultural historian Dr. Tomás Martinez Saldaña of the Colegio Postgraduado in Texcoco, south of Mexico City. The arrival of the new settlers from the south had ecological implications, not only changing the ecosystem but also forcing the Tewa to adjust economically.

Oñate signed a contract with the Spanish Crown in 1595. This contract can be called the first proposal for economic development in what was to become New Mexico. An inventory done in Santa Bárbara, Chihuahua (Española's sister city), in 1596 and 1597 documents what the Oñate settlers brought with them. Oñate had procured 312 *fanegas* of corn, some 12 fanegas of beans, and 500 fanegas of wheat seed, though most of these might have been consumed during the expedition's subsequent delay. A fanega is a dry measurement equal to twelve gallons. In the past, people measured plantings by fanegas instead of acreage. What did survive was a medicine box that contained beans, barley, and lentils, most likely in flour form, for plasters. Among the medicinal herbs brought by the settlers were *manzanilla* (chamomile), *eneldo* (dill), *ruda* (rue), *estafiate* (a Mexican herb), and *malva* (mallow). Domestic animals, which could provide food and also be used for breeding stock, included 846 goats, 198 oxen, 2,517 sheep, 383 rams, 96 colts, 101 mares, and 41 mules and jackasses.

The new settlers, in other words, introduced plants, animals, and tools that would quickly alter the landscape. Thus the ecology as well as the diet of the upper Río Grande was changed forever. Just as the Anglo-Saxon culture views the world through different lenses than the Indohispano culture today, so too did the new settlers (vecinos) and the Tewa see the world through different cultural glasses.

The Tewa, who were subsistence farmers, foragers, and hunters rather than herders, had learned to adjust to unexpected weather changes. Agriculturalists, though not immune to climatic change, can often regulate the ecosystem by raising crops in favorable locations and can alter nature using artificial irrigation and terracing. Their ecological relationships illustrate different strategies for survival that, over the course of four hundred years, have blended into one, though differences still persist even if the blood has mixed over time. Probably nowhere else is this a daily reality than in the foods eaten by both Indo-hispanos and Pueblo families.

According to the historian Gaspar Pérez de Villagrá's *History of New Mexico*, the new settlers gave the pueblos lettuce, cabbage, peas, garbanzos, cumin seed, carrots, turnips, garlic, onions, artichokes, radishes, and cucumbers. More than four hundred years later, these same crops are still grown in the area. Oñate and a group of advanced scouts arrived in Ohkay Owingeh, at the confluence of the Río Grande and Río Chama, on July 11, 1598, and Oñate reported, "On the 11th [of August] we began work on the irrigation ditch." The first thing the settlers did, even before the others arrived on August 18, was construct an acequia, for without water they couldn't do anything. This indicates that they intended to settle permanently.

Then, "on the 23rd [of August] the building of the church was started, and it was completed on Sept. 7." The blessing took place on September 8. The acequia was imperative for the wheat harvest the following year in order to replenish the settlers' depleted stores. The plants, animals, and tools—especially the iron ax—brought by the settlers would soon change the landscape. The following year the new neighbors demonstrated new technologies that the Indians would soon adopt: plowed fields, irrigated wheat, and kitchen gardens for growing vegetables and herbs. Kitchen gardens were a new innovation and their polyculture added variety to the Tewa diet, complementing the wild plants they gathered. Orchards were also established either as hedgerows or on the irrigated land.

A letter sent from San Gabriel in 1601 by Juan de Torquemado stated, "Irrigation water [from the acequia] was used for fields of wheat and barley and maize . . . and all other things that were planted in gardens because in that land are . . . cabbages, onions, lettuces, radishes and other small garden stuff . . . many good melons and *sandias* [watermelons] . . . wheat, maize, and Mexican chile all do well."

When Fray Benavides came through New Mexico in 1625 (he published his *Memorias* in 1630), lentils, habas (broad beans or favas), lima beans, and vetches were all growing and doing well. Benavides mentioned plums, peaches, and apricots specifically, but not apples. Benavides also observed, "So fertile is the land that it has been seen to harvest a 120 and a 130 fanegas to each fanega sown of wheat." When melons and watermelons came to the upper Río Grande is not known for certain. But their large fruits and sweet taste were symbols of a prosperous harvest.

In 1599 Oñate noted, "There are fine grape vines, rivers, and woods, with many oak and some cork trees; there are also fruits, melons, grapes, watermelons, Castilian plums, capulins, piñon, acorns, native nuts, *corolejo* which is a delicate fruit, and other wild plants. There are also many fine fish in this Río del Norte and other streams." In 1601 he wrote, "Our wheat has been sown and harvested; it does extremely well in that land. The Indians devote themselves willingly to its cultivation."

Besides the Mexican and European plants (many from North Africa and the Middle East), kitchen gardens (huertas) were introduced as a new method to grow vegetables. The new settlers also introduced a number of plants domesticated elsewhere in the Americas, among them chile and new varieties of corn such as the large-cob Cristalina de Chihuahua corn and the high-rowed Mexican dent from highland Mexico. This corn was more productive and was adopted by the Tewa. The settlers also brought Hubbard squash, known as calabaza mexicana, from South America, as revealed by seeds from Picurís. Also introduced was the nonfood plant tobacco, or punche as it is known in Spanish.

Herd animals became a double-edged sword; on the one hand they became a source of meat and textile material and served as beasts of burden. On the other hand, they overgrazed the grass, trampled the young trees, and compacted the soil. By 1601 the breeding stock had grown to three thousand sheep and cattle.

Probably what caused more ecological change was the metal ax, for it allowed for the felling of huge trees instead of only gathering dead limbs for fire. The same thing happened with the power saw, when it supplanted the ax.

On the plus side of what the new settlers brought, there was a several-fold increase in the number of domesticated plant species, which provided a more beneficial and secure subsistence base. Draft animals permitted easier access to distant sources of wood, and riding animals opened new hunting grounds. The new plants and animals safeguarded against ecological disaster, and the changes might have provided cultural continuity, in that they may have been responsible for preventing the settlers from migrating to another location. In a sense it anchored them to a specific site, where they have remained for over four hundred years, alongside their neighbors, although as they say in Spanish, "juntos pero no revultos" (together but not mixed), as each live in their separate communities.

The Embudo Land Grant

A Brief History

≈

THE LAWS OF THE INDIES
AND NEW MEXICO'S LAND GRANTS

LAND SETTLEMENT IN NEW MEXICO is a very complicated issue. Three documents in particular are critical for understanding how the land was settled after the arrival of the first settlers under the auspices of the Spanish Crown. The first is the "Ordenanzas de descubrimientos, nueva población, y pacificación de las Indias" promulgated by King Philip II in 1573. These ordinances were in effect when Juan de Oñate followed the Camino Real de Tierra Adentro from Zacatecas in 1596. The second is the ordinances that are most familiar to land grant scholars, activists, researchers, and historians: the Laws of the Indies of 1681. The third, issued about a century later, between 1783 and 1785, was the Plan of Pitic, written in what is now Hermosillo in the Mexican state of Sonora.

The birth of a land grant, or merced, can be compared to the birth of a child in that there is a specific date when the land grant came to be. I will use the Embudo grant as an example to illustrate such a "birth" based on the *cedulas* (royal decrees) of the time. For a land grant, conception corresponds to the first inquiry into the availability of a vacant piece of land, as in the case of Embudo de Picurís in the summer of 1725.

I will attempt to take readers step by step through the process of how land for a grant was obtained by settlers according to the Laws of the Indies of 1681, the law in force after the resettlement of New Mexico by don Diego de Vargas in 1692, following the Pueblo Revolt of 1680. The discussion references the Laws of the Indies by book, title, and law number, with the text of the law in italics. The text from the "Ordenanzas de descubrimientos, nueva población, y pacificación de las Indias" of 1573 is also included, with the ordinance number given in brackets. Most of the laws discussed here are those under Title 7, "Concerning the Settlement of Cities, Villages and Towns"; Title 12, "Concerning Sale, Composition and Apportionment of Lands, House-Lots, and Water"; and Title 17, "Concerning Public Roads, Lodging Houses, Markets, Inns, Boundaries, Pasture, Mountain Waters, Trees, and the Planting of Grape Vines," all in Book 4.

Before a particular area was selected for settlement, certain criteria had to be met. For that we go to Book 4, Title 7, Law 1:

> *The new settlements shall be established under the conditions of this law [Ordinances 39 and 40]. . . . In . . . inland settlements, the settlers shall choose the site from among those that are unoccupied, and may be occupied by Our order, without being prejudicial to the Indians or natives, unless it is with their free consent.*
>
> *When they make the plan of the place, they shall divide it into its squares, streets and house-lots, marked out with straight lines, starting from the main square and proceeding from it with the streets to the entrance and principal roads.*
>
> *They shall leave enough open area that, even if the settlement greatly increases, it will always be possible to follow the plan and expand in the same way.*
>
> *They shall try to have water close by so that it can be conducted to the town and properties, distributing it if possible, in order to make the best use of it.*

They shall try to have the materials that are needed for buildings, farmlands, cultivation, and pastures, so as to do away with the considerable labor and expenses that result when the materials are far away.

They shall not choose sites for settlement in places of very high elevation, because of troubles with the winds and the difficulty of service and transportation. Nor shall they choose sites in places of very low elevation because persons are apt to become ill.

Settlements shall be made in moderate elevations which benefit from exposure to the winds from the north and the south; and if there are mountains or hills, the settlements shall be on the east and west sides.

If high places cannot be avoided, they shall make the settlements in places where they are not subject to clouds, observing whatever is most conducive to health, and considering unforeseen circumstances that can occur.

And, in case of building on the bank of some river, they shall arrange the settlement in such a way that the sun may shine on the town before it shines on the water.

One can't find a better example of sustainability. The law focuses on finding everything locally, so as not to travel far for materials. It is a philosophy not simply of "subsisting," as scholars say the vecinos were, but of being sustainable. It appears that those that who applied for the lands in Embudo de Picurís were familiar with this law, for the land was "unoccupied" and it had enough "open area" and "water close by," with the Río del Pueblo (today Río Embudo) running through the center of the proposed settlement and the Río del Norte (Río Grande) forming the north and west boundaries. There was also plenty of land for "farmlands, cultivation and pastures"; and for settlements to be "made in moderate elevations" (the elevation of the grant ranges from approximately 6,200 to 5,800 feet) and of course, "health" was very important.

The historic Plazuela del Embudo, today known as Dixon, follows to the letter the law's dictate to "arrange the settlement in such a way that the sun may shine on the town before it shines on the water." Go to the

Plaza del Embudo when the sun comes out in the morning. Summer and winter, the sun first hits the old plaza and then the river because is it situated on the north bank of the Río Embudo.

The lands in Embudo de Picurís also adhered to Book 4, Title 7, Law 3: *"The ground and vicinity shall be abundant and healthful [Ordinance 111]. We order that the ground and vicinity that is to be settled shall be chosen in every way possible as being the most fertile, and abundant with pastures, firewood, timber, metals, fresh water, native people, viability for transportation, and accesses for entry and departure. There shall be no lagoons or swamps nearby in which poisonous animals might breed, or where there might be pollution of air or water."*

Again, the settlement in Embudo met all the above criteria, as the land is among the most fertile where all types of fruits, legumes, and vegetables can be grown. At one time there was plenty of pasture in the nonirrigated lands of the grant, with plenty of firewood in the Cañada del Oso, Cañada del Medio, and Cañada de la Orilla at the eastern end of the grant. The grant is also situated close to both Picurís, ten miles to the northeast, and Ohkay Owingeh, fifteen miles to the southwest. There is also plenty of "fresh water" from the Río Grande and Río Embudo, and the Camino Real de Tierra Adentro went through the center of the grant, providing "viability for transportation, and accesses for entry and departure." Above all, there are "no lagoons or swamps nearby" where "poisonous animals might breed" and of course no "pollution of air or water." Even today Embudo has some of the cleanest air and water, including the night sky.

Embudo de Picurís, or El Puesto del Embudo de Nuestro Señor San Antonio, as it was named by the new settlers, was settled in 1725 by a group led by three men from El Puesto de la Soledad del Río Arriba (present-day Alcalde) within the Sebastián Martín land grant, first awarded to Sebastián Martín (the older brother of Francisco Martín, who settled Embudo) in 1703 and later reissued in 1712. According to the petition made to the governor, the group was looking for new land for the expanding population, and they requested the land for agricultural purposes. The Embudo lands had been cultivated by the Picurís and Ohkay

Owingeh people prior to Oñate's arrival in 1598, as will be shown later. Above present-day El Bosque—on the north side of the Río Embudo, within the grant boundaries—there are ruins of an Indian pueblo. Oral history has identified at least five Indian settlements within the grant boundaries. The lands at Embudo de Picurís were supposed to have been (and still are) very fertile, producing all sorts of vegetables and fruits as well as a rich bounty of fish and eel from the Río Grande and Río Embudo.

Before applying for a piece of land, the petitioners had to be very familiar with the area they wanted. To use a sports analogy, they had to scout the area and also follow the royal requirements, in this case the Laws of the Indies. And one of those requirements was sufficient public lands. Book 4, Title 7, Law 13: *"Sufficient public land shall be designated for the town [Ordinance 129]. Public land shall be extensive enough that, if the settlement increases, there may always be enough space for the people to have recreation and for the livestock to graze, without causing damage."*

Since the vecinos (settlers) who settled in northern New Mexico didn't hit the jackpot when the Seven Cities of Gold became a mirage, they had to depend on the land to survive. Therefore, in selecting a site for their new home, pasture was essential as well as mandated by royal decree. Book 4, Title 7, Law 13:

Pastures and lands shall be designated for public use [Ordinance 130]. Having designated enough land for public use of the settlement and its growth, as has been ordered, those who have the authority to . . . establish a new settlement shall designate pastures that adjoin the public land for pasturing of work oxen, horses, livestock for butchering, and of the usual amount of other livestock that the settlers are ordered to have. They shall designate an additional proper amount of pastureland for the council. They shall designate the remaining lands as farmlands, which they shall apportion by a drawing of chances, and there shall be as many of them as there are house-lots in the settlement. If there are irrigated lands, they shall likewise be apportioned in the same way to the first settlers by lottery, and the rest of these lands shall remain unassigned, so that we may grant them to those who come to settle later.

Aware of the royal requirements, and knowing well the land they wanted to request, on July 17, 1725, the petitioners, Francisco Martín (brother of Sebastián Martín), Lázaro Córdoba, and Juan Marquez, residents of La Soledad del Río Arriba within the Sebastián Martín grant, went before Governor and Captain General Juan Domingo de Bustamante in Santa Fe and presented the following petition:

To the Governor and Captain General:

We, Juan Marquez, Francisco Martín and Lázaro de Córdoba, native citizens of this kingdom and residents of Río Arriba, present ourselves before your Excellency in that due form permitted us by the Royal Requirements and represent that whereas we are without any lands to cultivate to enable us to meet our obligations, whereby not only ourselves but our children (who are not few) and our wives, experience continual want year after year, and to the end that we may, God willing, supply those wants, we have agreed together to register a piece of vacant public land, and therefore royal domain, at the embudo de Picurís, so called outside the limits of said Pueblo three leagues more or less.

And from some fields cultivated by the natives of said pueblo outside of their lands as aforesaid, separated by a dry creek running from south to north forming the eastern boundary, we register up to the Río del Norte, which forms the western boundary, and on the South to the boundaries of Captain Sebastián Martín, and on the North the said Río del Norte, and though the Picurís river passes through these lands we petition for the lands lying on either side of the stream.

We therefore humbly request and pray your Excellency to grant us said land in the name of His Majesty (whom may God preserve) as we are poor and needy, whereby we will receive benefit and favor and justice, for all of which we pray and we swear before God and upon the Holy Cross that our petition is not made in dissimulation but in great necessity and declare whatever is necessary.

Juan Marquez
Francisco Martín
Lázaro de Córdova

Without doubt the royal requirements to which the original settlers alluded in their grant request are the Laws of the Indies of 1681. They qualified for a new grant, given that they had no "lands to cultivate," according to their petition. And if they did have land in La Soledad del Río Arriba, which more than likely they did, especially Francisco, as was the younger brother of Sebastián Martín, they would have to give up their claim to qualify for the grant under the law.

Book 4, Title 7, Law 18:

Declaration concerning which persons will go as settlers in a new colony, and how they are to be qualified [Ordinance 45]. We order that, when people leave any city, the Justicia and the government shall be obligated to have submitted before the notary of the council the qualifications of those persons who wish to go to make a new settlement. All married persons, and sons and descendants of settlers who do not have house-lots or lands for pasturing and farming in the place they are to leave shall be permitted. Those who do have shall not be permitted to leave because a place that is already settled shall not be vacated.

Governor Juan Domingo de Bustamante replied on July 17, the same day the petition was presented to him:

Before me, the Governor and Captain General of this Kingdom of New Mexico, came the petitioners with their petition, and viewing the same with an eye to justice I took the same into consideration, and in regard to what the parties ask I did and do command that Captain Miguel José de la Vega y Coca, Chief Alcalde of this City, be commissioned to proceed to the place mentioned by the parties making the petition presented to me and examine the land described.

And being without injury to any third party having a better right or to the native Indians, he will in the name of His Majesty (God preserve him) make the grant petitioned for under the circumstances they mention, and no person or persons having a better right appearing to contest the proceeding, he will place the parties in royal and personal possession, from which let them not be ejected without being first declared.

And let them settle the land within the term ordered and provided by His Majesty in the Royal Statutes, and let no authority impede said Chief Alcalde in carrying out this commission, for I do clothe him with the authority of Royal Law. Thus his Excellency decreed, commanded and signed, before me the present Secretary of Government and War.

Juan Domingo de Bustamante
By Command of His Excellency the
Governor and Captain General

Antonio de Yrrusisaga
Secretary of Government and War

Two days later, Captain Miguel José de la Vega y Coca, alcalde of Santa Fe, and ensign Miguel Enrriques traveled to Embudo de Picurís to examine the piece of land requested by the petitioners. They observed,

At this place, embudo de Picurís, on the nineteenth day of July, one thousand seven hundred and twenty-five, I, José Miguel de la Vega y Coca, Chief Alcalde of the town of Santa Fe, having exhibited the above decree, made by His Excellency the Governor and Capitan General of this Kingdom, to the ensign Miguel Enrriques, Chief Alcalde of said jurisdiction, he replied, placing the same upon his head, that he would obey it and that it may so appear he signed this with me on date aforesaid.

Miguel José de la Vega y Coca
Miguel Enrriques

What the petitioners didn't expect was a protest from the Picurís Indians. There was a provision, both in the 1573 Ordenanzas and the Laws of the Indies, that forbade the harming of Natives in the settlement of new places. Book 4, Title 12, Law 7: "*That the lands be divided without exception to people, and without harm to the Indians [Ordinance 136]. It is ordered that the division of lands, in new settlements, same as in places that are already settled are done with all justification, without admitting singularity, exception of persons, nor damage to the Indians."*

Captain Miguel José de la Vega y Coca responded to the Picurís:

And immediately thereupon, on said day, month and year, I, said Captain Miguel José de la Vega y Coca, Justice commissioned for these proceeding, proceeded to examine the tract mentioned by the petitioners in their petition, accompanied by the ensign Miguel Enrriques and the Indians of the Pueblo of San Lorenzo de Picurís.

And having seen and ascertained that they petitioned legally therefore, and it not being to the injury of said Indians who having heard and had explained to them the decree made by His Excellency the Governor and Captain General of this Kingdom, presented to me through their interpreter called Juan, versed in the Castilian language and in the presence of their Chief Alcalde and the witnesses in my attendance, who were the lieutenant Jacinto Martín and Miguel de Quintana, two papers, through which they claimed title to said tract.

One of them being General Juan Flores Mogollon, now with God, in which he directs Captain Juan Ruiz Cordero, Chief Alcalde of that day, to gather all the fire arms in the hands of said Indians, and the other showing the proceedings had by said Chief Alcalde relative to the removal of the Indians of the Tigua tribe who might be residing at the said Pueblos of Taos and Picurís thence and their settlement at the pueblo of San Agustín de la Isleta.

And finding that they had no title in the premises, gave to understand what the papers did say, and they conferred upon them no title whereupon they stated that they put their horses upon the tract to graze when they came to plant their fields. And this not being a good argument I asked them whether they had any fields on this registered tract and requiring them to point out the same to me, and they pointed out to me two of the best pieces of land and the same being uncultivated and uncleared, it is manifest they had never been fields.

And on explaining to them this fact they stated that their ancestors planted there at the time of the conquest by and when the first
Spaniards came into this kingdom in the time of Juan de Oñate.
Finding that all their arguments do not support or tend to support
their claims, I took from the tract a piece of land the petitioners had
registered, to see whether this step would satisfy the Indians, treat-
ing them as I did charitably and actuated as I was by a desire for the
quietude of all parties. But the result of this generous proceeding
was that they became arrogant and contended for more than ever,
whereby they clearly manifested the desire to prevent the settlement
of the Spaniards.

And without any reasons, more than those above mentioned, on
which account, and knowing as I do, and it being quite notorious that
since the conquest of this kingdom they are not known to have had
any field on this tract, or besides those they have had and now have in
use and cultivation above. And this grant not being injurious to
them, for the only injury they assert would result is, that if the place is
settled by the Spaniards their livestock will do them damage. To
which I, the said commissioned Justice as well as their Chief Alcalde,
replied causing them to understand that, should they be damaged,
they will have justice and will therefore be paid damages.

And in regard to their horses they have ample room for them in
the wooded hills, and which supply plenty of pasturage and thus for
all the reasons set forth, as well as for the fact known to me, that
these Indians have ample cultivable lands (with that at the Embudo
distant more than three leagues from their pueblo, and that which
they use surrounding their town) and as they raise ample produce
for their support.

I, by virtue of the authority conferred on me by his Excellency
the Governor and Captain General of this Kingdom, proceeded to
give royal and personal possession to the petitioners in the name of
His Majesty (God Preserve Him) which possession they received in
his royal name, proceeded by the ceremonies prescribed by law.

And I took them by hand and walked over the tract, and they
as upon their own property and in sign of possession, cast stones,

plucked up weeds and shouted, and have the same boundaries as those named in their petition and it is required of them that they settle the same within legal time. And it may so appear in these proceedings and in this royal grant, I said Captain Miguel José de la Vega y Coca, acting by appointment, signed the same, with the undersigned witnesses in my attendance, on said day, month and year.

Miguel José de la Vega y Coca
Jacinto Martín
Miguel de Quintana

Once the grant was official the settlers had three months to take possession of the land. Details of what had to be accomplished in those three months can be found in Book 4, Title 12, Law 11:

That possession be taken of divided lands, within three months, and it be planted, with penalty of losing them. All settlers who received land will be obligated within three months, after being assigned, to take possession and plant all boundaries and adjoining lands with willows and trees, and with time prepare all the soil and have it in good condition, some of which can be used to harvest firewood, that they maintain, with a penalty that if time should pass and nothing is planted, the land will be lost, so that it can be provided or given to some other settler, so that space be available in terms of land, but also in the village and ditches that are within every city or town.

First on their agenda, more than likely, was to lay out the acequia madre, for water was needed for everything. Then, once the acequia was constructed, they would plant, as was the law. And even before building a house, they would "plant trees," as recommended by Gabriel Alonso de Herrera in his *Obra de agricultura* of 1513 as well as in the Laws of the Indies.

Book 4, Title 17, Law 11, states:

The lands be irrigated according to this law. It is ordered, that the same order that the Indians had in the division and allocation of water be observed and practiced among the Spaniards where it has to be divided, and the land being marked, and for this involve the natives, that used to be in their charge, that they be watered the same, and each one gets the water, which they are entitled, successively from one to the other, with penalty for those who want priority and take it, taking the law onto their own hands, it should be taken away from them, until everyone has watered their field.

Though the law makes it very clear that the *pobladores* (settlers) should follow the Natives' tradition and custom ("It is ordered, that the same order that the Indians had in the division and allocation of waters be observed and practiced"), it also invokes the customs and traditions brought by the Moors to Andalusia in southern Spain: "and each one gets the water, which they are entitled, successively from one to the other, with penalty for those who want priority and take it, taking the law onto their own hands." In essence this law takes into account the customs and traditions of both Indians and Moors and creates a new set of laws that is truly New World.

Next, once they had dug the acequia and planting was started, the pobladores had to start building their houses. Book 4, Title 7, Law 15: "*Having planted, the settlers, should start building [Ordinance 132]. Once the planting has been done, and the livestock is in place, and with the help of God a good crop is expected, begin with a lot of care and diligence to build your houses with good foundations and walls, and start constructing retaining walls, garden plots, and have all tools and instruments needed for building fast and economical.*"

A provision regarding livestock is found in Book 4, Title 7, Law 26:

That the settlers plant, then take the livestock to the common lands, where they won't cause any damage to the Indians [Ordinance 131]. Then, and without delay, that the agricultural lands be divided, the settlers plant all the seeds that they had with them, and they should be prepared; and for greater facility that the Governor appoint a person to be in charge of planting the grains and vegetables to better supply themselves; and in the common lands keep all the livestock that is possible, with their brands and marks, so that they can start breeding and multiply, in places where they will be safe and won't cause any damage in the lands, planted fields or any other property belonging to the Indians.

The Picurís Indians raised concerns about the vecinos' livestock damaging their fields, which the settlers had been cautioned against in the above law (Ordinance 131). Also noted here is that the governor should appoint what we might call a mayordomo to be in charge of planting. This was also stipulated as part of the Plan of Pitic, that there should be one mayordomo for the acequia and another to make sure everything was planted.

Even to build a house, the settlers had to follow the royal decrees. More than likely, during the first few years the pobladores lived in *jacales* (temporary houses made of sticks and mud), though according to royal requirements all houses had to conform to a certain style, three hundred years before Santa Fe style was mandated by law. They had to be built in the form of a *plazuela*, to "serve as a defense and fortification." Book 4, Title 7, Law 17: *"That the houses be built according to his law [Ordinance 133]. The settlers agree, that the lots, buildings and houses follow a certain pattern, with the same adornments, and that they enjoy the north and south winds, uniting them, so that they can serve as a defense and fortification for those who would want to obstruct or infest and by all means all houses should have enough space for the horses and beasts of burden, with courtyards and pens as wide as possible, so that they be healthy and clean."*

While building and working, the settlers were to avoid contact with the Natives, though this might have been easier mandated than done. Whether that happened in practice is not known. Book 4, Title 7, Law 24:

That during work there is to be no communication with the Natives [Ordinance 137]. All while the new settlement is finished the settlers should at all possible avoid communications and contact with the Indians. Don't visit their pueblos, don't separate, or lose concentration while working the land, nor permit the Indians to enter into the inner circle of the population, until everything is complete and ready for defense. And the farm houses, when the Indians see them, would cause them admiration, and they understand that the Spanish are settled there permanently, and they fear and respect, so that they desire their friendship and not offend them.

And though each vecino (settler) had up to three suertes (long lots below the acequia), the dehesas (the commons) belonged to everyone. Book 4, Title 17, Law 5: "*That grazing, mountains, water and boundaries be in common. It is ordered, that the use of all grazing, mountains, and waters in the Provinces of the Indies be common to all the settlers, who are now settled, or will settle, so that they can enjoy freely.*"

Not only were all pastures, mountain lands, and water to be shared by all, but once the harvest was in, the stubble fields also became common lands. Book 4, Title 17, Law 6: "*That the land planted in grains, once harvested, they serve as common grazing [Ordinance 34]. The lands and property within the grant, or bought in the Indies, once harvested, remain common pasture lands.*" This was still practiced throughout northern New Mexico, including the Embudo grant, as late as the mid-1950s and even into the 1970s. Some people would cut the fences to allow their cattle onto the *rastrojos* (stubble fields).

Also held in common were the fruits that grew in the wild. Book 4, Title 17, Law 8: "*That the fruits in the mountains be common. Our desire is to make all wild fruits in the mountains common, and that each individual can harvest and take plants to transplant in their property or farm, and advantage should be taken being that they are common.*" Even today people harvest plants and trees from Forest Service and BLM lands because they consider them part of their heritage, and they don't see their actions as breaking the law, because of this particular law.

The laws cited above are those I consider the most important for attempting to understand how a land grant was made and what had to be done to maintain possession of the land. Also of paramount importance was water. Contrary to popular belief, the vecinos had to prove that the land they desired was vacant, that their possession of it would not infringe on the Indians who lived closed by, and that the land was fertile with a good climate. The law required them to observe the natives to see if they lived to old age and if they were healthy. The vecinos were also concerned about air and water pollution, and their houses were to be built on the lands above the irrigated fields.

If today we would follow the Laws of the Indies by not building on wetlands and keeping away from irrigated agricultural land, our landscape no doubt would be in better condition. The Laws of the Indies should be studied not only as a historical document but also as an organic, living document that can provide insight into today's development.

Francisco Martín, known as "El Ciego," appears to have been the leader of the three who received the Embudo grant, since most of today's native population can be traced to him. He was the brother of Sebastián Martín, who in 1703 had received the grant that bears his name. The two others mentioned on the Embudo grant papers are Lázaro Córdoba and Juan Marquez. I am a descendent of Francisco Martín. Part of his house survives in the plazuela, where it has been remodeled and is maintained in very good condition. The settlers were looking for a place with land and water for agriculture, livestock, and the growing of food. The areas that formed the grant had been used for agriculture, cultivated by both the Picurís and Ohkay Owingeh people; before Oñate ever made his way here in 1598 these lands were already providing a livelihood. According to local oral history, at one time there were five Indian "pueblitos" of people living here. One was above El Bosque, on a *banquito*, or natural terrace facing the south; another was east of the plazuela, northwest of the present-day school on the Martinez property. Supposedly there was one up the arroyo that divides Cañoncito and Montecito. My cousin José Agustín Arellano still remembers having seen the remains as a child. There is also archaeological evidence backed by oral history that at the mouth of the Cañada de los Comanches, where the Embudo empties onto the Río Grande, on the north side of the river, there used to be Indian vecinos. A flood in 1948, according to Teresa Archuleta, made these people flee their house, and the flooding river destroyed the sliver of garden land where they planted. Lands at the Embudo were supposed to have been, and still are, very fertile, growing all sorts of vegetables and fruits and providing a rich bounty of fish and eel from the cold waters of "el Bravo."

Though the grant was made in July, the pobladores didn't actually take possession until September 1725. When a new grant was made, the custom among the new settlers was to pull weeds, yell, shout, and celebrate in honor of their new home. In the case of the Embudo land grant, the triangular piece of land consisted of approximately twenty-five thousand acres. Today less than seven hundred acres are in private hands, with the Bureau of Land Management and the New Mexico State Land Office owning most of the grant land.

When a piece of vacant land was settled, first on the settlers' agenda was to lay out an acequia, an artificial means of moving water. Once that was constructed, they would plant, as was dictated by ordinance. And even before building a house they would plant trees, as recommended by Gabriel Alonso de Herrera. Undoubtedly the first acequia in the Embudo grant was La Acequia del Llano, since it brought water to the center of the plaza. It has its beginning in Montecito and empties into the Río Embudo at its desagüe, or sleuth, by the Arroyo de la Mina. The crescent-shaped piece of land that is irrigated by the Acequia del Llano had to have been the original rancho of the grant. This is the longest acequia and the one with the most parciantes. Analyzing the layout of the land and its proximity to the plaza, the next acequia to be built had to have been the Acequia de la Plaza.

THE SEBASTIÁN MARTÍN LAND GRANT: THE VELARDE TO ESPAÑOLA CORRIDOR

This area is mostly grassland above the acequias that dissect the land starting at the canyon in Velarde to Española on both sides of the Río Grande. The altitude ranges from 5,650 feet in Española to 5,800 feet at the mouth of the canyon. The rainfall here is about ten inches per year. Irrigation water is drawn from the Río Grande via nine ditches, known as the Nueve Acequias. Velarde at one time was known as La Joya, implying a very fertile land, as *joya* is a jewel. Though called La Joya, it should be Jolla—the hollows. Soils here are mostly deep, old alluvium with good fertility, but due to overgrazing, the lands east of Highway 68 are not in very good condition. Though there is still some grazing in the commons on the old Sebastian Martín land grant, there are also erosion problems due to the effect of off-road vehicles and four-wheelers. And now overcrowding from housing developments west of State Road 68 to the old Camino Real that runs above the acequias is compounding the erosion and waste problems.

Historically a lot of chile and corn as well as fruit was grown in this area, especially in Velarde or La Joya. San Juan or Ohkay Owingeh, which at one time produced a lot of corn and chile, today leaves most of its 1,800 acres of irrigated land fallow. The Velarde area still produces most of the apples in the valley, though a lot of the orchards have been cut down and have been taken over by Siberian elm. There are a few farmers who still plant several acres of chile, but not as much as before. Wineries are also doing a good business in the valley; there is one in Velarde and two in the Embudo area, and a wine grape co-op recently formed. This area seems to be good for grapes, and since grapes use less water than other crops and there is a demand for good wines, this might be a money crop for the future. Now a lot of the agricultural land is being subdivided for housing, mostly mobile homes, which is affecting the water quality of the area due to all the septic tanks that have to be installed. Especially troubling are those built close to the river in wetland areas, areas where historically people never constructed houses. If people were more aware of the historic ordinances the agricultural land wouldn't be in its present deteriorated condition.

Embudo de Picurís Watershed: The Physical Landscape

We have now traveled from the Indus Valley across North Africa to the Iberian Peninsula, across the Atlantic to the Americas, and from the sacred valley of the Incas to Mesoamerica. From there we followed the legendary trail-turned-road the Camino Real de Tierra Adentro, until we arrived in Ohkay Owingeh, north of the ancient villa of Santa Cruz de la

Cañada, the heart of the Española valley. From there we look east and glance at the jagged peaks today known as the Sangre de Cristo. In the distance we can see a bald peak known to those of us from the area as La Jicarita. This is the anchor, or the head of what is known as the Embudo watershed, which rises from an altitude of 5,800 feet, starting at present-day La Junta, where the Río Embudo empties onto the Río Grande, to a little over 13,100 feet at the north end of Truchas Peak. From Jicarita (Gourd Mountain), looking west, the watershed stretches south to the tip of Truchas Peak and north to the top of La Junta Canyon, then west to where the Río Embudo waters meet those of the Río Grande, born in the San Juan Mountains in southern Colorado. Below the north side of Truchas Peak are two lakes that feed the Trampas River, known as Laguna Escondida (Hidden Lake) and Trampas Lake. The creeks that converge to form Río Embudo are Trampas Creek, San Leonardo Creek, Santa Barbara Creek, Río Pueblo, La Junta Creek, Alamitos Creek, and Quemado Creek. San Leonardo Lake, Lake Alice, Lake Ruth, and Lake Hazel also feed the stream system. The watershed is in the form of a V, or an *embudo* (funnel); it also resembles a man on a cross, with Jicarita forming the head and its two arms stretching out to la Junta Canyon and Truchas Peak, and the feet where the two waters meet at La Junta de los Ríos, present-day Embudo.

The watershed contains three broad vegetation zones, with a fourth in the highest altitudes. In this area primitive conditions are still preserved. There are no roads, no commercial timber cutting is allowed, and no developed campgrounds exist. The area is open to public hunting and fishing; travel on foot or horseback is permitted, but no automobiles. Most of the human occupancy is along the streams, which feed the approximately forty acequias madres in the piñon-juniper and ponderosa pine zones. There are about 8,500 acres of arable land, with most of the acreage falling in the Peñasco area and only about 700 acres of irrigated land in the lower Embudo Valley (including Rinconada, which gets its water from the Río Grande), watered by ten historic acequias.

Piñon-Juniper Brush Zone

This zone, which is characterized by steep slopes, mesas and deep arroyos, canyons and cañadas with level or gently rolling terrain, stretches along the Río Pueblo, Río Santa Barbara, and Río Embudo at altitudes between 5,800 and 7,500 feet. Soils here are mostly deep old alluvium with good fertility, though erosion is a major concern over most of this zone, especially in the foothills neighboring the villages of upper Santa Barbara along the Río Trampas, Chamisal, and in the lower Embudo, in Cañoncito, Montecito, Apodaca, Bosque, Dixon, Embudo, and Rinconada, though this last village lies along the banks of the Río Grande. Though there hasn't been any grazing, off-road vehicles have made deep scars in the fragile hills throughout the valley and the arroyos have become eyesores, not to mention an environmental and health problem, as they have become dump sites. Average rainfall in the lower Embudo Valley is twelve inches per year, with the Trampas and Ojo Sarco area receiving fourteen inches.

Pine Zone

This zone encompasses the communities of Ojo Sarco, El Valle, Llano de San Juan, Llano de la Yegua, Rodarte, and Peñasco, stretching all the way to Tres Ritos. The terrain consists of rolling foothills with several steep slopes. Watershed conditions along the eastern boundary and in the northwest are generally satisfactory. The remainder of the area has poor plant cover and severe erosion, especially close to the villages. Private lands constitute about 88 percent of the pine zone in the national forest. In the Peñasco and Llano area the average precipitation is sixteen to twenty inches per year, while the Tres Ritos area gets about twenty-five inches yearly.

Spruce-Fir Zone

There are no human habitats within this area and the vegetation cover is mostly in good condition while the soils are somewhat stable. Erosion is not much of a problem. This area covers a lot of the Pecos Wilderness, the area around Jicarita Mountain, up the La Junta canyon, and south toward Jicarilla Peak. Average annual rainfall in this zone is around thirty inches.

Alpine Grasslands

This zone has the highest precipitation and water yield of the water- and foodshed. The locals call this area the water bank of the watershed, for this is where the snow is stored for use during the irrigation season that stretches from early April to November. Soils are mostly of sedimentary or metamorphic origin and the rate of plant growth is very slow. This area encompasses the upper reaches of Jicarilla Peak, the area around Hidden Lake and Trampas Lake, north of Truchas Peak, and the bald area of Jicarita Mountain. There is serious erosion on the denuded slopes due to runoff from melting snow and summer rains. This area receives approximately forty inches of rain per year.

The Upper Embudo Watershed:
Picurís, Peñasco, Chamisal, Trampas, Ojo Sarco

The headwaters of the Embudo Valley also create an embudo-shaped (funnel-shaped) watershed that forms natural terraces, descending from Santa Barbara to La Junta de los Ríos, extending from the western slopes of the Sangre de Cristo Mountains on the northwestern shoulder of La Jicarita and traversing down through the Alamitos. The boundaries extend to northern Cañon de la Junta, above Tres Ritos, and the Laguna Escondida that receives the water of the northwestern part of Truchas Peak, the eastern boundary of El Chimayoso and Jicarilla Peak to the south of the watershed. The geological embudo is formed between Vallecitos and Cuestecitas, where the Río Pueblo cascades past the Río Lucio to the east to meet the Río Santa Barbara, which is augmented by the Río Chiquito, slivering through the verdant vegas in Peñasco with a serpentlike motion to become Río Embudo where the Río Trampas meets the others.

Up above, on the north arm of La Jicarita, the Alamitos is rejuvenated by countless arroyos with water, in contrast to those further down the valleys that are typically dry, the Ríto del Cañon de la Junta and the Piedralumbre Creek in Tres Ritos. La Jicarita and Truchas Peaks are part of the Sangre de Cristo Mountains. These waters form the rivulet known as the Río del Pueblo, which cuts through the middle of Picurís Pueblo. Las Mochas, Placita, and Vadito also get their water from the Río del Pueblo. Rodarte, Santa Barbara, Llano Largo, Llano de la Yegua, Peñasco, and Río Lucío all irrigate with water from the Río Santa Barbara; so do Chamisal and Ojito. The Río Santa Barbara's waters originate on the west side of La Jicarita, a majestic reminder of where our water is stored and where it trickles down, enough only to never die of thirst or hunger.

Farther south, the villages of El Valle, Trampas, and Vallecitos get their water from the east side of Chimayoso Peak, the north side of Truchas Peak, and Jicarilla Peak, to give birth to the Río Trampas. This creek originates at Laguna Escondida, where *la laguna de arriba* and *la laguna de abajo* unload their winter stash to give birth to the fields down below. Below the Río Lucío the Río del Pueblo meets the Río Santa Barbara and, by present-day Cuestecitas, all three come together to form the Río Embudo. El Rito del Ojo Sarco, though it is a very minor tributary, also contributes some water to the Río Embudo, as do some *ojos* (springs) on the Cañada del Oso.

John Baxter wrote in *Dividing New Mexico's Water* that in 1755 the residents of Trampas "made plans to divert [the water from the Rito San Leonardo] for irrigation in the Cañada del Ojo Sarco, a proposal vigorously opposed by residents of Truchas, who claimed an exclusive right." Apparently the Truchas settlers had been promised the water by Governor Tomás Veléz Cachupín, since they had been given permission to construct an acequia two years before the grant was made. However, the people from Trampas were determined to expand farming into the Ojo Sarco portion of their grant. In 1836 they started a legal battle that resulted in the water from the Rito San Leonardo being declared free and available for use by members of the Trampas community in Ojo Sarco. There were threats of violence but the Trampero irrigators prevailed.

HUMANS IN THE LANDSCAPE:
THE MORA-PICURÍS WATER CHALLENGES

Some of the water that used to flow west now flows east to the Mora Valley, due to an agreement between Picurís Pueblo and the people of Mora starting in the 1830s. The water starts at an elevation above 13,000 feet, but by the time it reaches the Río Grande at La Junta de los Ríos the elevation has dropped to 5,800 feet. Today expansion in the Tres Ritos area (which includes a ski resort) threatens not only the water supply but also water quality in these ancient communities, though for some the ski area means employment.

Today little is left to show how farmers on the other side of La Jicarita accomplished the engineering feat of taking water from the Río Grande watershed and transferring it into the Canadian River watershed. The first acequia that diverted the water to the Canadian River side was dug in 1832, with the permission of the Picurís Indians who were the first inhabitants of the watershed. But as the population grew, more water was needed, and therefore more acequias had to be dug to take water from the west side of the watershed to the east side. Not much has been written about this remarkable accomplishment, one attained by people of the land who at times used an old whiskey bottle filled with water as a level, according to an account by Therese Griffiths and Laura Robertson titled "The Flow of Mountain Water," which appeared in *New Mexico* magazine, March 1979. There is also a more scholarly account by the historian Dr. Anselmo Arellano of Las Vegas, "Acequias de la Sierra, and Early Agriculture of the Mora Valley" (1994), which can be found at the Santa Fe Public Library.

In July 1882 Governor Juan Pando of Picurís Pueblo filed a complaint on behalf of his people in Taos District Court (a copy of which is in the state archives at Santa Fe). Named as defendants in the complaint were Migual García and twenty-two other residents from the Agua Negra area. Soldiers from Fort Union were called in to intervene and several people were supposedly killed. Eusebio Arellano, who was eighty-seven years old when Griffiths and Robertson interviewed him in 1979, "recalled his father's saying that soldiers from Fort Union set up camp at Peñasco and intervened to stop the fighting. 'Some people were killed,' said Arellano. 'They are buried up there on the mountain.'" The question left unanswered is who were these people? Were they from Picurís? From Peñasco? Or, from Agua Negra (today Holman)? And did people really die?

The *New Mexico* magazine article continues,

Today only two of the ditches still serve their communities as well as they did on the days of their completion: Acequia de la Sierra which supplies Holman (Agua Negra), Cleveland and Mora, and La Presa Sierra Ditch, which supplies Chacón, with the rest flowing on to Holman.

The Acequia de la Sierra is born at Jicarita Peak and the high mountain valleys along the northern edge of the Pecos Wilderness in the Río Grande watershed. The ditch carries water across the mountain divide to the east, and spills it into the upper creeks of the Mora River. About three miles downhill from the tiny settlement of Medina the acequia separates into three channels—one to Holman, one to Cleveland and another short one angling off to the northeast.

La Presa Sierra Ditch draws from the headwaters of the Rito la Presa, as well as natural southern drainage, and drops in a spectacular waterfall into Griego Canyon above the town of Chacón.

In a document at the New Mexico State Archives, an eyewitness has left us this account:

The Mora people took the water of the middle branch [of the headwaters of the Río Pueblo—one of the three main streams that make up the Río Embudo] many years ago and they took the water of the northern branch 15 to 20 years ago. During the last three years the people of the little known town of Agua Negra have been building a ditch to the southern branch and last April 1882 the water was turned into the new acequia. It was constructed by about 14 men who were

provisioned by Padre Jean Baptiste Guerin, Parish Priest living at
Santa Gertrudes [in Mora].

These were built to move water between the Río Grande and the
Canadian watersheds, thus giving the settlers in Agua Negra
(Holman) and Chacón additional water for farming. This was no
easy task, often taking several years to accomplish. They were built
with the very crude tools and instruments of the time. Three were
built with the last being completed in April 1882.

In "Acequias de la Sierra and Early Agriculture of the Mora Valley,"
(1994), Dr. Anselmo Arellano writes,

[A] resident who lived on Picurís Pueblo lands was Antonio Olguín,
a soldier who played a major role in the early settlement of the Mora
Valley. The Hispanic population near Picurís continued to expand
after 1800; and finally, in 1816, Olguín and a group of families needing
agricultural lands and water set aside their perilous fear of Indian
attacks and engaged in a new settlement venture. They traversed the
Jicarilla Mountain and descended into the fertile Mora Valley on
the eastern border of the Rockies. This effort consequently launched
the settlement of the northeastern sector of the Spanish frontier in
New Mexico.

But they encountered a problem: the Mora River did not carry sufficient water to meet the needs of the growing population. Arellano goes
on to say,

Olguín, the early leader of the San Antonio settlement, also rallied
the people to confront the emerging problem of an inadequate supply
of irrigation water from the Mora River. He approached the Picurís
Indians and successfully requested permission to take some pueblo
water from the high mountain valleys and the crest of the Jicarilla
Mountain. The water was to be diverted from the western watershed
whose tributaries followed a natural course into the Río Grande. The
plan included an ingenious scheme to cut an irrigation canal into the

rock and across the mountain into the Mora Valley. The water taken by Olguín and the settlers flowed into one of three branches of the Río Pueblo that irrigated the farmlands lying within the Picurís Pueblo land grant.

The acequia was connected to the middle branch of the headwaters of the Río Pueblo. Through hard labor and native ingenuity, the people were able to defy the gravitational flow of water in places, elevating it until they created a major diversion into the Mora Valley along the eastern watershed. Although the exact date that the acequia was constructed is unknown, testimony provided in 1882 stated that the water from the middle branch of the Río Pueblo was taken "many years ago . . . by the individual Antonio Olguín . . . [who] was allowed to take this water." In view of this evidence, the acequia had to be constructed before 1832, since Olguín did not return to San Antonio with the families who resettled the valley in 1835.

According to Lorraine Aguilar's genealogical research, there was a José Antonio Olguín who was born in 1769 and died on December 4, 1835; in 1791 he married a María Dorotea Garcia de Noriega in San Lorenzo de Picurís.

The original families who settled the Mora Valley were largely from Las Trampas, Chamisal, present-day Peñasco, Embudo, and even Santa Cruz. It appears that prior to 1865 the people from the Chacón area began construction on another mountain acequia, taking water from the northern branch of the Río Pueblo for their own use. This acequia was referred to as the Acequia de El Rito y La Sierra. Today residents call it the Acequia de la Presa y la Sierra. It begins at a holding dam built at the top of the mountain, on the Junta Canyon. From there the acequia snakes around the mountain and releases the water into the valley starting in early spring.

In 1879 plans were made for yet another transmountain acequia, which would take water from the third and final branch of the Río Pueblo, the southern branch. The parish priest at Santa Gertrudis, Juan Bautista Guerin, met with the residents of Agua Negra and they agreed on fourteen men to do the work. Writes Arellano, "When they reached the stream, they also built a dam across it to hold the water which would be diverted to the fields at Agua Negra. After three years of arduous, backbreaking toil, the acequia was completed; and on April 1882, the water was turned into the new 'Acequia de la Sierra.'"

Statehood

Toward the end of the territorial period, in 1907 New Mexico's water code was revamped. Though the 1907 Territorial Water Code recognizes acequias as enjoying a distinct class of water rights protected by the Treaty of Guadalupe Hidalgo and governed according to Spanish and Mexican water law and local custom, it also allows for the separation of water from the land. Prior to 1907 water could not be sold separately from the land; in essence the new code made water a commodity. In 1912, when New Mexico joined the Union, the New Mexico Constitution confirmed all preexisting water rights.

Santa Barbara Tie & Pole Co.

The Santa Barbara Tie & Pole Co., organized by A. B. McGaffery from Vermont in 1907 to cut timber for the Santa Fe Railway, was the first company to enter and start cutting on what had been land grant property. Between 1909 and 1926 the Santa Barbara Tie & Pole Co. harvested over sixty-five thousand acres of Forest Service land. A special train, No. 3, was constructed by Lima Locomotive Works in 1909 and it was used in Hodges until the early 1920s. In New England ties were floated down the rivers, so McGaffery did the same in the Embudo watershed, using the Río Embudo. Some were picked up at the present-day Embudo Station Restaurant, when it served as a train depot for the Chile Line train that ran from Antonito, Colorado, to Española, New Mexico, between 1881 and 1942. Other ties were floated all the way to Cochiti. For a while this provided employment for local men from the Embudo and Peñasco area, but it also destroyed a lot of the habitat along the banks of the Río Embudo, especially around present-day Dixon.

The Chile Line and the Fiesta de Santa Rosa

The Denver and Río Grande Western Railroad was affectionately called the Chile Line because of all the chile and fruit that it carried from the Española Valley to the San Luis Valley. The first train to use the line supposedly arrived at the Embudo depot on August 30, 1881, the day of Santa Rosa, and a big fiesta was held. The community has celebrated the Fiesta de Santa Rosa ever since. For the past few years the Fiesta de Santa Rosa has been an annual community fiesta with food, music, and a parade. In the past dances were held at such places as the Sala Filantropica, the Sala Mutua in the center of town, and Tres Palomas Bar about one hundred yards west. Partygoers would go from one sala to the next. The Gracias Troupe from San Antonio, Texas, known locally as "los maromeros," also performed. People in the valley were notified of the event by "Sacando el Gallo," a form of town crier.

Española was founded in the 1880s as a stop on the Denver and Río Grande. From 1887 until the line's abandonment in 1941, passenger service in Española was generally daily except Sunday. The railroad has disappeared, but the city has grown and prospered as the commercial center for the valley.

Los Alamos

Two years after the Chile Line became history, in 1943, the "Secret City" on top of the Pajarito Mesa was born. More than anything else since Oñate and his settlers arrived in the Española Valley, Los Alamos had profound effects on the land and water in the valley. Prior to the establishment of Los Alamos, most of the villagers in the Española Valley and the Embudo watershed survived off of the land by maintaining their acequias or working on the railroad or as sheepherders in Colorado, Utah, and Wyoming. With the advent of Los Alamos, most of the men returned to their villages and became wage earners, though as the lowest-paid employees since most were uneducated. As their earning power increased and Los Alamos grew, Española became the hub for all the surrounding villages. The first food store, Fairview Foods, was established in the midfifties. Of course there were "mercantile stores," such as one in what was then San Juan Pueblo and the Bond and Willard where the ill-fated Plaza de Española is now located. The first fast-food establishment, Lota Burger, was set up in the 1960s. And as Española grew as a result of the expansion in Los Alamos, fewer and fewer people tended their farms and acequias. In one generation the people in the valley went from a pastoral economy to a postindustrial economy, bypassing the industrial epoch almost completely. Their only contact with the industrial epoch was the short-lived romance with the Chile Line.

This area of the upper Río Embudo historically grew mostly grains, especially wheat, and hay for pasture for livestock. Up to the Depression, wheat and two varieties of local corn, *maíz concho* and *maíz de los rincones*, were grown mostly for human consumption, either fresh on the cob or saved for winter use as posole and chicos. These two local varieties of corn have now practically disappeared. Chicos are made from corn that is harvested while tender, in the state called *xilote*, then cooked overnight in the husk in adobe ovens known as *hornos*. In the morning when the corn is pulled out of the oven, the husk is pulled back and the ears are tied up by the leaves and hung to dry in *ristras*. When the corn is dry it is removed from the cob and cleaned so it is ready to cook. This area also produced peas that were consumed either green or mature and dry in soups. Haba beans also do well here, though today they are hard to find. Calabazas mexicanas, which are eaten either green, while tender, or when ripe in pies or alone, grow well at this altitude, since they don't get the bugs that destroy the pumpkins and squash in the lower valley. Potatoes also grow well here and can be grown as high as Tres Ritos, though only one family grows them today. Potatoes and cabbage were grown as high as Arellano Canyon going up to La Junta Canyon. Today most of the land is used for cut hay or pasture to feed the few cows that remain. Historically most families had a few head of cows, some sheep and goats, one to three pigs, and up to fifteen or twenty chickens. Today very few people have any domestic animals. In the fifties there was a small dairy in Vadito.

The villages in the Upper Embudo start at Tres Ritos on the Río Pueblo, then Sipapu Ski area in what was known as Los Mochas, then Placita, Vadito, and Picurís Pueblo. There are also houses scattered

Horno, making chicos, Embudo. Photograph by the author.

starting at the Mora-Taos county line, an area known as the Mondragones. On the Río Santa Barbara the first houses are up by Hodges, then there's Santa Barbara, today named Llano Largo. Above, on the north ridge, is Llano de la Yegua, and on the south, Llano de San Juan. Down below is Rodarte. At the center of the valley the major economic town is Peñasco and then Río Lucio. Going south one comes upon the town of Chamisal; southeast is Ojito and west Vallecitos. On the Trampas, above is El Valle, then Trampas situated on the main road between Truchas and Peñasco. Over the ridge is Diamante, then west is Ojo Sarco, and the last hamlet before descending towards Cañoncito is Cuestecitas, named thus by the late Chester Salazar, according to his dad, Silas, who lived there until his death at the age of over one hundred.

LAND GRANTS IN THE EMBUDO WATERSHED

For our purposes, the most important mercedes, or land grants, within the Embudo watershed and the present-day Española Valley are the Sebastián Martín, Embudo, Santa Bárbara, and Trampas. Those who settled on these grants are all descendents of Hernán Martín Serrano, a native of Zacatecas who was forty years old when he made the trek up the Camino Real with Oñate in 1598. Of all the settlers who traveled with Oñate, the Martín Serrano clan undoubtedly made the biggest impact in the Española Valley. The four land grants that they settled on in the valley were squeezed between Ohkay Owingeh on the south and Picurís on the northeast.

The Martín Serranos considered themselves indios, according to scholars. It is said that once two Martín Serrano first cousins wanted to marry and the church wouldn't allow them. They responded, "We are Indians and don't have to follow church law," and they got married.

Ousted during the Pueblo Revolt, the Martín Serranos returned to Santa Cruz de la Cañada and present-day Los Luceros after the reconquest by de Vargas in 1692. It's here that Sebastián Martín, the most famous of all the clan, was awarded the Sebastián Martín land grant in

La Junta, Embudo Grant. Photograph by the author.

1703, a grant that was reissued in 1712. The grant went from the boundary with Ohkay Owingeh all the way to La Joya (present-day Velarde) and extended east to Ojo Sarco, including what became the Trampas grant.

In 1725 the Embudo grant was given to Francisco "El Ciego" Martín, Sebastián's younger brother. Its boundary on the south was the Sebastián Martín grant, and to the east was the dry arroyo before one climbs up to present-day Cuestecitas. The Río del Norte, today the Río Grande, formed the north and west boundaries.

In 1751 the Trampas grant was carved out of the Sebastián Martín grant. Sebastián may have given up part of his grant to a group from the Barrio de Analco, the Tlaxcalan settlement in Santa Fe, because they had married some of the original Martín settlers. In 1796 the Santa Bárbara grant was awarded to Valentín Martín, grandson of Francisco. By 1800 all of the lands between Ohkay Owingeh and Picurís were controlled in one way or another by members of the Martín Serrano clan. Today the most prominent name in the area is Martinez, and many who are not named Martinez have a mother or grandmother with the name. Within the Embudo grant most, if not all, of the acequias were constructed by members of the Martín clan.

All the lands within these land grants were divided according to the Laws of the Indies. The villages were all started with a plaza as the centerpiece and a church at one end of the plaza. Land was then divided among the settlers, with the suertes (long lots) allotted based on luck, or suerte, thus the name. Eventually these suertes were cleared of piñon, *sabina* (alligator juniper), and *chamiso*, and these tierras muertas (dead lands) were revived. The altitos (highlands) carved out and planted in orchards and vegetables, while the joyas or jollas (hollows) were broken down into *melgas* and then further compartmentalized into *eras* (sunken beds), huertas (vegetable gardens), *huertos* (orchards), and milpas (cornfields). All such land divisions were water-conservation strategies that will be elaborated on later. The uneven lands, where it was practical, were turned into different types of terrazas (terraces)—*bancos* on slopes, *bancales* in valleys, and *ancones* along the river—with *azoteas*, small, flat-roof-type beds, usually close to the house. In addition to being a strategy for water conservation, these terraced divisions prevent soil erosion and thus maintain the fertility of the soil, or what is known as the *flor de la tierra*, "virginity of the soil."

Besides suertes, each settler family was granted a *solar*—a site to build a house. A solar usually included a *dispensa* (utility room) and eventually also contained the *leña* (wood pile), *común* (outhouse), *trochil* (pigpen), *gallinero* (chicken house), and *corrales* (corrals), though these structures could also be in the commons. Since the solar was customarily on nonirrigated lands, either above the acequia or below on a piece of tierra muerta, or marginal land, it had little impact on agricultural lands.

Though in the Acequia Junta y Ciénaga most of the solares today have encroached on the irrigated portions of the suertes (especially in La Ciénaga), that was not the case before, mainly because there were fewer houses and the suertes were mostly intact. Today every time a house, garage, or shed is built, agricultural land is taken out of production, especially if the old philosophy, which has served us so well for four hundred years in New Mexico, is not followed. Still today in Spain and Mexico you hardly see any development on agricultural lands. In the Junta y Ciénaga of the early 1900s there were only about nine houses; today the same area has fifty-six houses and mobile homes. Problems especially occur when buildings and parking spaces are not laid out with what is known as *curia*—that is, creatively.

The Embudo post office is now located across the Arroyo Jacinto that separates La Junta de los Ríos (the juncture of the rivers), the place where I live, and La Ciénaga (wetlands). There were actually ranchos, or *granjas*, located south of the Río Grande and Río Embudo and west from El Puesto del Embudo de Nuestro Señor San Antonio (where today the Dixon post office is located); some of the old documents also referred to the new settlement as Embudo de Picurís, more than likely because the lands of Embudo were at one time planted by the people of Picurís, although they were also utilized by those of Ohkay Owingeh.

THE EMBUDO WATERSHED ACEQUIAS

The Upper Watershed

There are close to forty acequias in the Upper Embudo watershed, which comprises the Río Pueblo, Río Santa Barbara, Río Chiquito, Río Trampas, and Ojo Sarco Creek. In the Río Pueblo, the uppermost acequia is at Los Alamitos, close to where the road descends toward Mora, in Mora County. Excluding the irrigated lands of Picurís Pueblo, the nine acequias on the Río Pueblo irrigate 785 acres. They are the Acequia de los Alamitos, Spring Ditch, Acequia los Mochas, Acequia Vadito North, Acequia Lower Vadito South, Acequia de Placitas del Sur Vadito, Acequia de la Otra Banda, Acequia de Leña Pesada, and Acequia del Pueblo de Picurís.

Acequias on the Río Santa Barbara irrigate 4,541 acres. There are thirteen major acequias, not counting the laterals or secundarias. They are the Acequia del Norte de Río Lucio, Acequia del Medio de Río Lucio, Acequia del Sur de Río Lucio, Acequia Madre de Peñasco, Acequia de

Abrieu, Acequiecita de Peñasco, Acequia de Peñasco del Camino, Acequia Sur de Rodarte, Acequia Madre de Santa Barbara, Acequia Madre del Llano Largo, Acequia Madre del Llano de San Juan, Acequia del Cañon Chamisal-Ojito, and Acequia del Llano de la Yegua.

The Río Trampas irrigates 585 acres and it includes the following acequias: Acequia Sur de las Trampas, Acequia Norte de las Trampas, Acequia Abajo del Valle, Acequia Arriba del Valle, Acequia del Llano de San Miguel, and Acequia de Ojo Sarco, which alone irrigates 200 acres. This acequia gets its water from the Rito de San Leandro that flows into the Trampas, then is diverted and runs for twelve miles before arriving in Ojo Sarco.

The Lower Watershed

The acequias appear to have originally watered one granja, or rancho, made up of several suertes. An acequia can also be seen as watering one terrace, though each suerte might be broken up across more than one terrace. Up to 1950 usually one family, or an extended family, owned most of the land watered by one acequia. An example is the Acequia de la Sancochada, where the land was owned by the family of Juan Isidoro Martinez and Albinita Maes de Martinez. Albinita, who lived to be over one hundred years old, was a full-blooded Apache, according to my father; they were his great-grandparents. They had twelve children, among them Ricardo, Manuel, Ramon, Rafael, Juanita, and María de la Luz, who married José Ignacio Arellano at the age of thirteen. My grandfather, José Agustín Arellano, was their eldest child, born in 1868.

Since the coming of the "hippies" starting in 1968 a lot has changed, especially how the acequias are viewed and maintained. In the Embudo grant there are ten major acequias that get their water from the Río Embudo. On the south side of the river, starting east by the box canyon, are the Acequia de la Sancochada, which waters mostly Cañoncito; the Acequia del Medio, which begins between the Arroyo de los Pinos Reales and the Arroyito del Agua and irrigates parts of Cañoncito and Montecito; and the Acequia del Llano and Acequia de la Plaza, which water the area starting at the Arroyo de Lorenzo to Angostura, also known as Las Pasaditas, where the Acequia Junta y Ciénaga begins its journey along the north side of La Mesita. Present-day Embudo (actually La Junta de los Ríos and La Ciénaga), where the Acequia Junta y Ciénaga is situated, is located approximately twenty-five miles southwest of the town of Taos and twenty miles northeast of Española, on the southeastern tip of Río Arriba County. This acequia waters approximately eighty acres, the ranchos of La Junta and La Ciénaga, and at one time (until about 1998) it irrigated the lands at La Nasa with the *sobrante* (excess).

On the north side of the river is the Acequia Arellano-Martinez, also known as the Acequia de Leonardo Martinez, which is the smallest, irrigating about thirty acres, but it includes the biggest apple orchard in the area. The Acequia de los Duranes, Acequia de la Apodaca, and Acequia del Bosque irrigate the piece of land extending from the north end of the river in Cañoncito, Apodaca (which means cranberry in Basque), to the plaza on the north side of the creek, to where the Arroyo del Pino, also known as the Arroyo de la Vaca, empties onto the Río Embudo. And the last is the Acequia del Rincón, where the two rivers meet. Then there are two very small acequias that irrigate at most seven acres, Acequia de las Pasaditas and Acequia de Eliseo Martinez, the latter of which hasn't been used in about twenty-five years. These two acequias don't have senior water rights since they were constructed in the 1940s.

Drawing water from the Río Grande is the Acequia de la Rinconada, which now gets its water through plastic pipes directly from the river instead of through the traditional presa, the same as the Acequia de la Bolsa. On the opposite side of the river is the Acequia de la Otra Banda. On the north side of Río Grande is the Acequia de los Roybales, which now irrigates at most ten acres and at one time watered about twenty acres of crops.

Though no official records have been uncovered as to which is the oldest acequia within the Embudo land grant, migration patterns and genealogy can be used to piece together the history of the acequias. Since most of the settlers came from the Santa Cruz de la Cañada and

Sebastián Martín land grants near present-day Alcalde, more than likely the first acequia was the Acequia Junta y Ciénaga if the settlers entered through the north side of grant, by the Río Grande. Settlements usually started at the mouth of a river, then moved upstream, but since the Camino Real was on the south side of the mesa the first acequia had to be the one that delivered water to the historic Plaza del Embudo, the original settlement. The first acequia to draw water from the Río Embudo, starting at the mouth of the river, was the Acequia de la Nasa, a few feet from the confluence with the Río Grande. Records recently uncovered date this acequia to 1783; later it got its water as sobrante, or excess, from the Acequia Junta y Ciénaga due to problems in maintaining its presa (dam). A diagram from 1895 at the BLM office in Taos shows two acequias where the Ciénaga meets the Río Grande, about five hundred feet west from the Junta y Ciénaga desagüe. All of the other acequias within the grant have only one name.

Alternatively, if there was no irrigation going on in the area at the time the grant was made in 1725, then the first acequia was possibly El Llano, since this acequia delivered water to the center of the plaza, where even as late as the 1930s there was an orchard where the Catholic church parking lot now stands. The remains of the acequia that took water all the way to the center of town are still visible.

But since the Picurís protested the granting of the Embudo Grant, claiming the land as theirs and saying that the Picurís and Ohkay Owingeh people already used the Embudo lands for cultivation, there might have already been irrigation. There are also records that indicate the lands in Rinconada had already been planted since the late 1600s, according to people from the area, though I have not seen such documentation.

As far as the other acequias, based on genealogy and settlement patterns it appears that the last to be built were the Leonardo Martinez, Sancochada (which means parboiled or half done, perhaps because it was constructed haphazardly), Medio, and Duranes. And why is that? Because all those lands were, up until about fifty years ago, settled by descendents of Juan Isidro Martinez (a descendent of Francisco Martín) and Alvina Maes Martinez.

Leonardo Martinez was the son of Manuel Martinez, son of Juan Isidro and Alvina. Fidel Martinez, who owned big chunks of land in El Medio and Duranes, was the son of Ricardo, another son of Juan Isidro. The Arellanos, who also owned a lot of land in Sancochada and El Medio, were also descendents of Juan Isidro by his daughter, María de la Luz Martinez, who married José Ignacio Arellano. All the Martinezes with land in Sancochada, Medio, and Duranes acequias were descendents of Juan Isidro, according to the genealogical research of Lorraine Aguilar from El Valle in the Trampas land grant.

Therefore, it appears that Apodaca, Bosque, and Llano were established earlier than the Martinez-Arellano, Duranes, Sancochada, and Medio acequias. Following the migration pattern on the west side of the Sangre de Cristos, acequias were established starting from the bottom of the river and moved up. Those of the east side of the Sangre de Cristos, in the Mora Valley, again based on migration, were started from the top of the watershed and moved toward the bottom.

The "merced del Embudo," known today as the Embudo Valley, consists of the communities of Cañoncito and Montecito on the eastern part of the merced, on the south side of the Río Embudo, four miles east of the Plaza del Embudo (present-day Dixon). On the northeast side of the river are Apodaca and El Bosque, known simply as "la otra banda," the other side, or as El Bosque de los Angeles. Southwest from El Bosque is the Plaza del Embudo. Directly across the *pareja* (above El Bosque), which served as a track for horse races in the past, and along the south side of the Río Grande is present-day Rinconada, known briefly in the 1880s as Durazno, probably due to all the peaches grown there. Today some of the best peaches are still grown there. West of Rinconada is La Bolsa. North of La Junta, between the Río Embudo and the Río Grande, is El Rincón. West of the plaza is Las Pasaditas, historically known as La Angostura, where the presa of the Acequia Junta y Ciénaga is located. Nearby used to be a *tunelito* through which the water traveled on carved rock, until it was destroyed by a highway department project in 1948, according to local people. Half a mile west is La Junta, which is separated from La Ciénaga by the Arroyo Jacinto. Two miles west of La Ciénaga is La

Nasa, the westernmost community in the Embudo Valley. La Nasa irrigates with the sobrante from the Acequia Junta y Ciénaga. Due to the Chile Line train station and recently the Embudo Station Restaurant, on the north side of the river, maps list this area as Embudo instead of La Nasa. This causes confusion about where the real Embudo and Dixon are, while Dixon is also mistaken for the Dixon Apple Farm in Cochiti, since a lot of apples are also grown there.

In 1744, sixty-four years after the Pueblo Revolt of 1680 and nineteen years after the Embudo grant was made, there was a mini Pueblo uprising. Embudo was at the crossroads, where the Picurís, Taos, San Juan (Ohkay Owingeh), San Ildefonso, and Santa Clara peoples met to plot the fate of the settlers. But after six years, in 1750, most of the original pobladores came back to stay permanently. The following year Nicolas Apodaca bought a piece of land from Manuel Martín and the hamlet of Apodaca was born on the north side of the river and the east side of the Arroyo del Plomo, also known as the Arroyo de la Apodaca. There are still families in the valley (mine included), whose roots date back to 1725. Embudo, Abiquiu, and Ojo Caliente are considered *pueblos genízaros*, or settlements where non-Pueblo Indians (mostly Plains Indians) settled, and their principal language was Spanish.

After nearly a century of peace—since Mexican Independence and the Grito de Dolores didn't have any effect on the life of the paisanos— the American invasion of 1847 changed their lives forever. There was a fierce battle, known as the Battle of Embudo, during the Taos Rebellion, in which over fifty paisanos—*mexicanos e indios*—lost their lives on the Camino Real de Tierra Adentro between La Joya (today Velarde) and La Plaza del Embudo, on what's known as the Cañada de las Entrañas. The *descansos* (memorials) are still etched on the basalt rocks; there are over fifty descansos engraved on the lava rocks on the south side of La Mesita, the exact location given on an army map of the battle, before the spot where it meets the Cañada del Embudo, also known as the Arroyo de la Mina, which connected the north to the south. The people of Embudo, with their handmade tools for tilling the soil, made of oak and piñon, were no match for the cannons carried by the Americans. According to a personal account by Rafael Chacon, the people of Embudo had to hide in the canyon above Cañoncito with their animals, though there was over two feet of snow.

But through all the turmoil, the ranchos kept expanding as acequias were dug and new lands were opened up for cultivation. Until the early part of the twentieth century the majority of the people lived off the land. Now they have catapulted to the computer age, but their roots are still anchored in the land and the acequias. Although the old agriculture is giving way to organic farming, sustainable agriculture, and perma-culture, along with drip irrigation, these are simply new ways of pack-aging traditional agriculture.

This area historically grew wheat, corn, and chile, with fruit grown mostly in the Rinconada area. The people, similar to those in the up-per watershed, also had a few domestic animals for household con-sumption and grew all types of garden vegetables. At one time one individual had about five hundred cattle that he grazed in the Tusas area by Tres Piedras. Though people grew mostly for home consump-tion, they would also peddle their chile and fruit in the Taos, Peñasco, and Mora Valleys and as far north as Questa, Costilla, and the San Luis Valley in southern Colorado. There they would sell for cash or barter (*cambalache*) for pinto beans, bolita beans, habas, dried peas, potatoes, and meat, mostly sheep. Today there are very few cattle and sheep left. Most people grow only small kitchen gardens for home con-sumption or for the farmers markets in Santa Fe, Taos, Pojoaque, Los Alamos, and Española and the Dixon Market. Just as the demograph-ics of the area have changed dramatically in the past thirty-five years, people grow mostly greens today, with very little chile or corn and no grains at all. At most only about 10 percent of the land is planted in small kitchen gardens. Almost everybody filled their properties with apple trees in the fifties, but a very cold winter when the temperature dropped to thirty-six degrees below zero on January 7, 1971, killed most of the orchards. As a result most of the trees were cut down for firewood and the few fruit trees that remain are only for home use or sales at the farmers markets. There are only three big orchards left, one in Cañoncito with twenty-five acres, another in Embudo with ten acres, and one in Rinconada with about twenty acres. A new fruit crop grown here commercially is the grape. The vineyards in Cañoncito are the farthest north in New Mexico, located at six thousand feet, and there are two wineries in the area and a brewery.

4

La merced

El juicio de la tierra

≈

THE ANATOMY OF A LAND GRANT

LAND PATTERNS IN NEW MEXICO trace back to the first type of land settlements under the Spanish Crown. From 1598 to the Pueblo Revolt of 1680 the land that was typically given to settlers was known as an *estancia*, and there were two types—*de ganado mayor* (for cattle) and *de ganado menor* (usually for sheep and goats) (land grants didn't come until after 1692). These types of land settlements are what today we would call ranchos, since jacales, temporary houses made of upright posts plastered with mud, were about the only structures that were allowed to be built on them. From 1692, following don Diego de Vargas's return to northern New Mexico, through the arrival of the third wave of settlers in 1695, the estancia structure was abandoned and mercedes, or land grants, were established where settlements and structures were permitted. Whereas the estancias were mainly huge expanses for livestock with no irrigation, the land grants, especially the Spanish grants made between 1695 and 1821, were more for agriculture. There were also *mercedes de agua*, or free distribution of water for irrigation, but these were not as common. The Spanish land grants were not as expansive as those made later under the Mexican government, between 1821 and 1847. Though there's been a lot of "chatter" regarding land grants since Reyes Tijerina's infamous 1967 Tierra Amarilla Courthouse Raid in northern Río Arriba County, very few people understand the land grants and the struggle of the people on them.

The topic of "traditional" agriculture in northern New Mexico is a lot more complex than what appears on the surface. The philosophy of the land grants depended on understanding *el juicio de la tierra*, or "the wisdom of the land." But traditional agriculture can never be understood without fully comprehending the division of the land based on the mercedes. We will look at the three main components of a merced. First, the land known as the commons in Spanish, called *ejidos*; second, the acequias; and third, the suertes that are irrigated by the acequias. The rigid design of the acequias separates the first component from the third like an X-Acto knife cutting a zigzag line across a paper. The suertes dot the landscape in such a zigzag pattern because they were allotted to the vecinos, or settlers, by lottery or luck. Each acequia in essence forms a separate terrace. If the suertes are the body, the acequias are the veins that give life to the high-desert landscape, and this in turn produces a holistic food. When you have a nourishing diet based on grass-fed meat from the commons, fruits and vegetables watered with fresh stream water that has been exposed to the sun, asparagus and *verdolagas* (purslane) from the *jardin rizo*, beans grown on *secano* (land used for dry farming), then your life is abundant. Dr. Tomás Atencio calls this life "una vida buen y sana" (a good and healthy life), to which I would add, *y alegre* (and joyful), and Reverend Antonio Medina from Mora adds, *y sostenible* (sustainable).

The *Ejidos* or Commons: The Acequia Water Bank

"You cannot deny your parents. You cannot deny your history, your roots." Thus admonishes the late Ricardo Legorreta, Mexican architect, in a December 21, 2005, article in the *New York Times* headlined "A Modern Space in Mexico City's Historic Center." For material objects were not the only items that made their way from Spain to Mexico and then to New Mexico. Probably the most important immaterial entities that traveled from the south to the north were ideas and philosophies as to how people related to land and water use in a new environment, though one that was very similar to that of the Iberian Peninsula. The system of land and water use braids together a number of elements, and to understand it one has to unravel the *trenza*, or braid, one strand at a time.

When Legorreta talks about not denying your parents, history, and roots he is referring to the Arab influence that reaches back to the Spanish colonizers' Moorish past. To most historians the so-called New World presents a black-and-white dichotomy of Spanish (Castilian) versus indigenous, or Mesoamerican, influence. And even here, Mayas and Aztecs take most of the credit, while the Tlaxcaltecas who came to New Mexico with the early Spanish settlers are not even mentioned, though they settled in Santa Fe, probably as early as 1600, and also around present-day Albuquerque, at Atlixco as in the Atrisco land grant. But history is not that simple; there are lots of shades of gray in the palette. It is only recently that the Sephardic, or crypto-Jewish, tradition has begun to be studied. What for all practical purposes is not even mentioned in scholarship is the Muslim influence, though about a third of all Spanish words are derived from Arabic. When analyzing land patterns in New Mexico we always go back to the Laws of the Indies of 1681, which are based on the Ordenanzas of King Philip II of 1573. But as when one peels an onion, when we start searching for legal antecedents we encounter the Arab influence in all aspects of land and water use in New Mexico, albeit under the guise of Roman law. The Moors are the prodigal sons.

Under the Laws of the Indies, the land was divided into what we know today simply as commons and irrigated lands. What divides the one from the other is a rigid zigzag line formed by the acequia, the channel that delivers the water and gives life to all the land below it. This rigid design line follows the contours of the land. Above the acequia is the dry land, which is more in tune with how the land was managed in Northern Europe prior to the arrival of the Arabs in the Iberian Peninsula in 711. When the Moors were kicked out of Spain, their methods for managing the land did not disappear; in fact, they resurfaced in the "Indies" under the guise of different ordenanzas, the laws under which the Spanish land grants were made to settlers.

When referring to the commons, many people in the Río Arriba region think of ejidos, which simply supplanted the word *latifundia* here. And though the term *land grant* has a high recognition level among the general population, especially Indo-hispanos, very few understand its anatomy. Latifundias are big expanses of land, in the thousands of acres, whereas *minifundias* are small landholdings of only a few acres. And *ejido* simply means "exitus," or the place at the outskirts of a village that is neither planted nor worked and is common to all. *Ejido* comes from the Latin verb *exeo, exis*, "to exit," "to leave."

There are four main divisions within an ejido, although they blend into and overlap each other at times, again like a braid:

- sierras
- montes
- dehesas
- solares

Sierras provided the early settlers—and still today the descendents of these early pobladores, like their ancestors before them—a place to harvest firewood as well as *vigas* (beams) and *latillas* (split pieces of cedar that go across the beams) for constructing houses and other buildings that needed to be built for survival. When the mercedes were awarded, building materials for living quarters were dragged from the sierra and monte using animal power; today trucks are employed for this type of labor. The settlers also combed the lands for wild fruits, *capulín* (chokecherries), *chatacow* (elderberries), *moras silvestres de matas y de suelo* (wild raspberries, alpine strawberries), piñon, and *beyotas* (acorns). Wild herbs, such as *oshá, oregano de la sierra y del campo, altamisa, poleo*, and *yerba buena*, were harvested, as they are today. Each village has its place where certain essential herbs are grown and harvested; many of these sites are kept secret. Since the coming of the flower children or hippies, many of these sites have been raided to the point of near extinction as some started harvesting the herbs to sell commercially. Capitalism, which started to creep into the Indo-hispano villages with the arrival of Stephen Kearney in 1846, finally arrived in full force in the 1980s, with the blossoming of artists' studio tours in practically every village. It's during this time that the *remedios de la gente*, (the medicinal herbs of the people) started to be sold in specialty stores in Taos and Santa Fe. What had always been harvested communally became the individual property of those who befriended the trusting *curanderas* to learn about the special spots that only they knew and then razed them for individual profits.

Curanderas, who in the past readily told others where they got their remedios, today are cautious about divulging their secret gathering places, whether in the high sierras, the juniper and piñon montes, or the high-desert dehesas that produce the *chimajá* (wild parsley) in the spring. Those familiar with the language know that most landscapes are named to signify where certain raw materials are found. For example, *el llanito del zacate de la escoba* meant "broom grass grew there," while *el arroyo del barro* identified the site as a clay deposit. Place-names also related to the local environment. Costilla, for example, is named for mountains that looked like ribs, and Questa signified going up or down the side of the mountain, *Embudo* means funnel, and the name was applied because the watershed takes the form of a funnel.

Like the allocated lands, communal lands or ejidos were broken down into sierras, montes, dehesas, and solares where the houses were built. But the commons were also crisscrossed by cañadas and *veredas*. A cañada can be described as a *camino mesteño*, "wild road," since it was used to move livestock, mostly sheep and goats, from winter to summer pastures and vice versa, from the dehesas to the sierras. A cañada is usually defined as a space between two high peaks, or *lomas* and *cuchillas* (mountain ridges), that has water holes, or abrevaderos, and vegetation for animals to eat and is at least ninety varas wide. Besides abrevaderos, cañadas also have spaces where the livestock can rest, called *descansaderos* or *majaderos*, which refer not only to a resting place but also to a place where manure is deposited. Also part of the commons are the veredas, or trails, which are more narrow but usually a minimum of twenty-five varas wide and usually used for horses or to move smaller flocks or herds of livestock. There is a *dicho*, or saying, "Quien deja el camino real por la vereda, piensa atajar y rodea" (He who leaves the royal road for the trail thinks he will make a shortcut but instead makes the road longer). Both cañadas and veredas are common roads. It's from the cañadas reales that the term *dehesa* might have originated, according to some scholars, since conflicts arose between those moving livestock and the inhabitants of the villages through which the animals were moved twice a year. So *dehesa* is thought to have derived from the term *defendere*, which means permission, since the king had to intercede and grant permission for the livestock to be moved. All of these concepts eventually made their way into the Laws of the Indies and thus to New Mexico.

Sierra is mountainous terrain. The term may derive from Arabic, referring to a rugged high desert. In Spain the word applies to high, sawtooth mountains, and it was appropriately transferred to the southwestern ranges by colonists who came under the Spanish Crown. It's in the sierra that the *cuencas*, watersheds, form, and the sierras act as the keepers of the water because the snow melts slowly. Thus these lands not only provide the irrigation water for the acequias but also feed the aquifers that feed the norias (another Arabic word, from *na'ura*), or wells, that provide water for domestic use.

Monte is derived from the Latin *mons* or *montis*, meaning *tierra alta*, "high ground," while *montaña* is *tierra alta, áspera y habitata*, that is, "highland, harsh but habitable." A verse about mountains says,

Preñado dicen que estoy	Pregnant they say I am
Y jamas a parir vengo,	But I come not to give birth,
Lomos y cabeza tengo	Loin and a head I have
Y aunque vestido no estoy	And though dressed I am not
Muy grandes faldas mantengo.	Huge skirts I have.

The mountains are said to be pregnant because of their huge *rumores y hinchazones*, the swellings and bulges that make them up.

Tienen también cabeza	They also have a head
Y es su cima y	And it's the summit and
espaldas	shoulders
Y sus vertientes llamamos	And the slopes that we call
faldas	skirts
Aunque no ande vestido,	Even though not dressed,
Y dicen comunmente	And are commonly called
Faldas de un monte.	The skirts of the mountain.

Mountains also have *cejas*, or eyebrows. The English language is not that precise in naming features of the environment, since people's names are more commonly used. An example is Pedernal Peak by Abiquiu that people wanted to rename O'Keefe Peak for the renowned artist Georgia O'Keefe. She had more sense and put a stop to that before she died. *Pedernal* means flint, and there's a lot of flint in the area.

The nonirrigated lands of the mercedes, especially those lands known as secano, used for dry farming, are usually on the lower reaches of the dehesas, known as *tierras de pasteo*, or pasture lands. In Latin the dehesa is called *pascua*, and it is a place where livestock are grazed. The term could very well come from the Roman custom of establishing latifundias in marginal lands. The term *dehesa* first appeared in the year 924 in the dictionary of Corominas, though Romans noted it could be found in the laws of the Visigoths. According to Covarrubias *dehesa* is an Arab term that means "a low land, full of weeds where it is hard to walk, from the moisture in the soil and thick with weeds." He explains that the word comes from *dehisetum*, from the verb *dehesa*, "que vale espesar y estrechar." But he adds that it could also be Jewish, from *dese, herba*, for the deshesa is nothing more than "a piece of land full of weeds." A dehesa is a seminatural ecosystem where there is usually a certain amount of human involvement. In New Mexico this meant that the piñon trees were pruned to the extent of removing what is known as *piñon blanco*, the dead piñon branches that have gotten a gray patina and were treasured by the ladies when they relied on firewood for cooking and heating, for it is seasoned wood. This type of piñon tree also usually produces the best piñon nuts and, because it has been taken care of, the nuts are easier to harvest.

A dehesa is also a space that conserves a great number of both flora and fauna; it also has great economic and social importance. Regardless of its original meaning, whether Latin, Arabic, or Hebrew, it is understood to be an agroforestal system with poor soil and a harsh climate where humans have intervened to make it somewhat productive. Some scholars say that dehesas are not ecologically sound due to economic pressures to graze more livestock than they can sustain. The Bureau of Land Management, the New Mexico State Land Office, and the Forest Service now manage dehesas, which once formed part of the different land grants. It is usually a type of pasture with scattered trees of evergreen piñon and juniper (*cedro y sabina*) and deciduous oak, and in the past grains were often grown under the sparse tree covers. The space between the dehesa and the solar, situated above the rigid line made by the acequia, which separated the commons from the private lands, was used for dry farming. A dehesa can be better understood as a mosaic because of its different uses. It's part monte but is also used for grazing and, when necessary, dry farming. The best pinto and bolita beans are grown on secano. It's an agroforestry system with the joint production of trees, agricultural crops, and animals; it's also known as an agrosilvopastoral system.

In addition to the sierras, montes, and dehesas there are the solares where the houses were commonly built. Solares are private and to a certain extent part of suertes, and they are usually located above the acequia. The term *solar* comes from the word *suelo*, "to make a floor," as in constructing a house on a plot of ground. But a solar is usually the space between the acequia and the commons. The house, if located away from the town plaza, was constructed in an "L" or "U" shape, like Moorish houses on the alquerías. Part of the house complex also included the dispensa, or utility room, and the *soterrano*, or root cellar, where people kept their food supplies for winter.

Rome conquered Spain in the fourth century BC, and thus Roman culture had a vast influence on everything where the Spanish settled, including agriculture and especially land use. The Romans introduced aqueducts that transported water to areas that were without water before, primarily in urban areas, although the technology also affected agriculture. They also introduced the plow.

The land itself dictates what can grow in a certain parcel of land; that is the wisdom of the land. The land grant landscape tells you when to plant and what will grow. Not only can food be produced where there is no water for irrigation, but certain crops do better with little water. Frijoles, or beans, produce better without any irrigation. This is known as temporal or secano agriculture. "Temporal" implies planting right before the rainy season begin, so that when the rains arrive the seeds are in the ground and ready to sprout.

Acequias: The Arteries That Deliver the Water

Después de regar, el agua deja
  su huella
	After Irrigating, Water Leaves
	  Its Mark

Anoche cuando dormía	Last night when I slept

Di, ¿por qué acequia
  escondida,
agua, vienes hasta mí,
manantial de nueva vida
de donde nunca bebí?
	Where, through which
	  hidden acequia,
	water, you come to me,
	spring of new life
	from where I never drank?

ANTONIO MACHADO

All of a sudden acequias have become sexy! They're the new superstar on the block, but like Hollywood's superstars, they are having problems dealing with their newfound fame. Don't say anything negative about the acequias or you will get a shovel blow to the head. Even some of those who work with acequias are being referred to as "acequia superstars," something that probably has the late Andres Martinez of Taos, Cleofes Vigil of San Cristobal, and Pablo Romero of Dixon (the old Plaza del Embudo) turning over in their graves. These men, who truly personified what it meant to work with acequias as mayordomos, who understood the system, never really got recognized for their work, and some of them even shunned the spotlight. They are the faceless and nameless individuals who made the bioregion flourish. The only thing they got for their acequia work and dedication over lifetimes that spanned seventy-five years or more were calluses on their hands. The little I know about acequias I learned not from books but from such people, including my father. You will never find their names, not even in footnotes, but it's because of them and many others throughout the Río Arriba bioregion that the acequias are still around. Since they left us the acequias have fallen into disrepair even though they now have acquired that exotic allure that makes them sexy, Hollywood-style.

Therefore, the purpose and urgency of this work is to try to convey the reason for maintaining the acequias, and hopefully retain some of the knowledge our ancestors had of how this carbon-free—since they operate on gravity—artificial, man-made system of utilizing water can be integrated into our environment before we destroy the acequias in the name of development. They are our querencia. We can love them to death. That is, if we don't understand how they function, we'll neglect them because we think that being good stewards of the land means doing nothing, like I was once told by an artist when we were cutting down a tree that impeded the flow of water in our acequia. If we get a tumor in our bodies, do we treat it or do we allow it to grow, because it's an organic growth? Acequias are organic systems like the human body, and if we don't take care of them we will accelerate their demise.

On the other hand, romanticism is killing the acequias, literally choking them to death. People not familiar with acequias might think that bank easements full of invasive species such as Russian olives, Siberian elms, and poison ivy are just part of the landscape, but that is not the case if you are the mayordomo who has to walk, or try to walk, along the banks. This type of neglect breeds chaos for the acequia easements, though it puts money in the bank of artists who prefer this type of acequia to paint or draw. This type of neglected easement puts money in their bank!

Back in 2001 our acequia had been so neglected that the width had closed to about fifteen inches, which led J. P. Lujan, a first-term commissioner at that time, to quip, "Not even a chicken can dance in the middle of the acequia." A beaver chose our acequia for her condo, and since the mayordomo didn't find the den in time, the acequia went to hell and overflowed overnight, while the beaver was having her grand opening. The whole side of the mountain came tumbling down onto the road. But it turned out to be a blessing in disguise, because it made us realize how badly maintained the whole system was and that the only way to approach it was to do some planning. Before, everything was done using the Band-Aid approach. For example, if there was a gopher hole that couldn't be plugged but there was no way to find it due to the growth along the banks, put in a culvert, *por mientras*, as we say in Spanish, "for the time being." Another leak, another culvert; but when not installed properly, culverts do more harm than good. If they are not leveled when they are installed the water backs up and, more important, the vegetation on the bank dies off.

I was drafted to be mayordomo, and I accepted on one condition: that we do a five- and a ten-year plan. Without realizing it, we were ahead of the curve. Now the state legislature is asking that acequias submit five-year plans called Infrastructure Capital Improvement Plans (ICIPs) to get funding. In 2006, when the heavy rains came, they destroyed not only our acequia but also several others in the valley, including the Acequia del Bosque, Acequia del Llano, and Acequia de la Plaza. My mom used to say, "Dios les ayuda a los huevones" (God helps the lazy), and I guess she was right. Due to destruction caused by the arroyos, which are a big problem I'll address later, the Federal Emergency Management Agency (FEMA) declared the acequias a disaster area and offered financial help. Again, *de pura chiripada*, "by a stroke of luck," under the five-year plan we were able to redo part of the acequia that needed a lot of work. We could have gotten more help but one of the commissioners said that we didn't want to get too greedy. I guess the military needs the money in Iraq to destroy their acequias.

But there's hope if people are willing to pay attention and at the same time disregard the notion that acequias are romantic landscapes created for artists to make money painting them. Acequias were constructed to bring water to communities, not only for domestic purposes but to grow food for survival, including water for animals. And they help recharge the aquifer from which we get our well water. There is a whole lexicon about acequias that is almost gone. This has to be relearned and taught. Once we get to the point of "historic preservation," that will signal the death of the acequias. It will be time for the archaeologists and anthropologists to write acequias' obituary and for musicians and poets to write romantically about them, since they will have been relegated to memory and will no longer serve their intended purpose, to provide life for the community that created them.

The community needs acequia literacy, for the majority of parciantes, realtors, judges, police officers are illiterate when it comes to the most valuable natural resources communities have. I have written that newcomers don't understand acequias, as some try to keep the mayordomo and the peones away from their property, not realizing that acequias have easements that can't be violated. Others try to pump directly from the river so they can have water when they want it; this, of course, will destroy the communal nature of the acequias. Some have suggested that instead of blaming the state of the acequias on newcomers, I should try to educate them. Let me clarify that it's not only newcomers who don't understand the acequias; it's also the native-born who have not learned about the acequias. They have never heard of the Ordinances of 1573 or the Laws of the Indies of 1681 or the Plan of Pitic of 1789, documents that describe the role of acequias in the layout of towns and villages. But it's not their fault either; it's the educational system that has failed us.

ACEQUIA DEMOCRACY

Acequias, it is said, are the most democratic institutions in this country, and it's true. An acequia is far more democratic than a voting precinct because it represents far fewer people, and the fewer people involved, the more democratic an institution. Take, for example, the Embudo land grant. There are two precincts that encompass all the voters living in the area—Democrats and Republicans, Greens, Independents, and so on. But this same land grant is home to sixteen acequias, most of which are historic acequias—that is, they predate the 1907 water code.

Each acequia that has at least four parciantes, or water-rights owners, has to have bylaws, a three-member elected commission, and a mayor-domo, who is also usually elected by the parciantes. Because people don't use their land for agriculture anymore, as the villages are becoming bedroom communities, some acequias have dropped by the wayside. In Embudo the Acequia de la Nasa, which watered with the sobrante, or excess water, of the Acequia Junta y Ciénaga, hasn't had water is several years. The Acequia de la Bolsa also abandoned its traditional point of diversion, as did the Acequia de la Rinconada, and now the property owners pump water directly from the Río Grande. These latter two acequias are also less democratic now, because they are no longer traditional acequias. The rituals have fallen by the wayside.

What ultimately makes the acequias more democratic than any other institution is that they share the water to the last drop. This concept, called *equidad*, or equality, comes directly from the Qur'an. Under Muslim law, possibly because it evolved in the desert, people must never deny water to another being. This is called the Law of Thirst. To portion water out to other beings, including animals and plants, is considered a *limosna piadosa (zakat)*, a "pious charity." This concept has been practiced in northern New Mexico forever. I am told that my grandfather always had a trough full of water for travelers and their animals. Our people never thought of selling water. Also, there was always a plate of hot food on the wood stove known as the *plato del pasajero*, or the "traveler's plate," since there weren't any restaurants then. This is part of the concept of convite. Today that is no longer the case.

Another aspect of the democracy of water is that it could never be severed from the land, because water was considered the blood that gave life to the land. There is a Spanish *refrán* that says, *El agua es la sangre de la tierra* (Water is the blood of the land). Water was always shared based on the amount of land one had. That's where the concept of peones (measures of water) comes in. A peón (in some acequias known as *tiempo*) can be broken down into quarters; usually a quarter peón meant the person had one acre of land to irrigate, and it was divided based on the twenty-four-hour day. Normally a quarter would allow them six hours for irrigation, but if the water had to be shared under repartimiento, that six hours might be reduced to as little as fifteen minutes. The apportionment also depended on the number of parciantes in a particular acequia; it was based (or should be) on the amount of land irrigated by the acequia.

Two other concepts besides repartimiento promote the democratic nature of the culture of the acequias: convite, from *convivium*, relates to food, and *cooperación* relates to labor. Acequias involve cooperation in the true sense of the word, for an acequia is a worker-owned co-op. In 1997 a man from the Mondragon Cooperative in the Basque region of Spain told me, "No puede haber cooperativas sin cooperación" (There can be no cooperatives without cooperation), a very simple concept but one that is difficult to implement in this country of individualism.

In the Hispano-Arabic world water does not belong to any one person or institution; it has to be shared equally among those who need it. The person in charge of administering the water, making sure everyone has water and doesn't abuse it, is known in New Mexico as the mayordomo and in other parts of the world as the *sahib al-saqiya* (the *zabacequia* or *aguadero, repartidor del agua*). Of course, the water was divided based on time and volume: the amount of water in the river, the amount of land each acequia irrigated, and the number of water users in each particular acequia.

An acequia also promotes food democracy based on the concept of convite. When my mom would prepare a special plate, she would tell me to go take a plate to my aunt or some special person, saying, "Dile que aquí le convido aunque sea un poquito" (Tell her that I am sharing even if only a small portion). Probably the ultimate convite philosophy was the *güeso guisandero*, a bone that was shared in times of very scarce resources to at least give the taste of meat to a gravy or seasoned dish. This bone, it is said, was passed from house to house.

Then there's the democracy of labor, whether it's the annual spring cleaning, or the repair of an acequia after a flood to get the water flowing again, or helping out neighbors during planting or harvesting. Just as the water is shared, so too is the labor, and those who have more land need more labor; some might help with labor and others might take a plate of food to the workers. This means that when the harvest is in, everyone will also partake of the harvest, whether it's with chicos (made of tender corn when it's in the xilote stage) made in the horno, a piece of meat after the *matanza* (the ritual of butchering an animal that usually started after the Fiesta of San Martín, November 11), or for Lent, a special bowl of *panocha* (a sweet desert made of ground *harina de trigo enraizada*, dried wheat sprouts).

An acequia, then, is the epitome of democracy, whether it be in the way in which the commission and mayordomo are elected, based on one vote per person regardless of whether that person has one acre or twenty acres; food allocation in which those who have provide for those who have not; or labor, where everyone cooperates, from the spring cleaning to putting away the harvest. Three words that define acequia democracy are *repartimiento*, *convite*, and *cooperación*. When any one of them is lacking, democracy begins to deteriorate.

THE ACEQUIA AS AN OASIS CULTURE

My knowledge of acequias comes from lifelong experience as a small farmer in Embudo. It hasn't come from books. My mentors were my father, whom I considered an expert on acequia irrigation, and others like him, old-time *labradores-sembradores* (farmers), mayordomos, and commissioners with a wealth of information. Since the time I was a little kid I would attend the annual acequia meetings with my dad. I was the only kid in a kitchen full of men smoking their roll-your-own Prince

Albert cigarettes, or *punche mexicano*, using tobacco they grew themselves.

After I became an adult I started doing oral history interviews. One from whom I learned volumes was the late Pablo Romero, who was mayordomo of the Acequia del Llano in the Embudo Valley. I owe the design of my land to him. He probably knew more about acequias than any other person I ever talked to, but he never allowed me to record our conversation. I couldn't even take out a pen to take notes unless it was a word I had never heard before; then I would have to ask him for a special dispensation to write it down. He never allowed me to photograph him either. So I would spend countless hours under his apple tree talking about acequias and the acequia landscape. But like an ethnographer, once I left his place I would pull out a folded paper I always carry in my shirt pocket for my notes and jot down the most important points. I was like Carlos Castaneda with Don Juan: if I wanted to learn, it had to sink in; I couldn't use a tape recorder or camera.

Others who have provided a lot of knowledge that I wouldn't have otherwise been able to obtain are the pícaros, those rogue characters who used to abound in the community but now are endangered themselves. From these characters I learned who has what seeds, who's a good cook, who invites you to lunch if you happen to arrive close to noon—anything dealing with agriculture they know. When I would pose a question to one of these pícaros, they would always say, "Curre con fulano de tal él te puede ayudar" (Go with so-and-so; he can help with your question). One thing I learned about a pícaro is never ask him to help you plant, because plant he will, the entire village if you want, but once he entices you to plant, he won't be back to help water, hoe, or do any of the *quehaceres* (chores) associated with farming. The next time you'll see him is during harvest time. Then he'll invite his friends and take credit, saying, "Este es nuestro jardin" (This is our garden). One might say the pícaro is related to what my mom would call *las visitas de agosto*, the "perennial August visitors": those who never visit in the spring, when there is work to be done, but who are the first to arrive once the harvest starts, like the swallows of Capistrano.

ACEQUIA DESIGN

The first question a person might have is: How did people know where to construct the acequia when there were no landscape architects, engineers, or hydrologists to guide them? They relied on men like my mentors, who were experts when it came to their landscape. The first thing they did before starting to dig an acequia was to survey the river for *veneros* (water springs), *venitas* (veins of water), and manantiales or ojos, springs or the origins of a source of flowing water. Of course, usually before people asked the Crown for a land grant, they had already become familiar with the land they wanted, and that included the water source. Once the source of water was identified, the next step was to survey the landscape and try to visualize the largest amount of land that the acequia would be able to water. Keep in mind that the landscape did not look at all how it looks today. It was covered with grasses, piñon and juniper trees, and other vegetation. The arroyos ran naturally. The landscape was virgin, in other words.

The acequias, it must be remembered, are man-made, artificial environments. That's why it bothers me so much when I look at the acequia landscape today and see that the present owners—both newcomers and local natives—are allowing the land to be covered with Siberian elm, Russian olive, and tamarisk and simply letting the land be destroyed, sometimes due to neglect and other times due to ignorance and their romantic vision of the landscape. Just as people have memory through their personal and family oral history, so does the landscape, in the form of scars that can be seen in the land, such as the different forms of terraces and strategies for conservation known as melgas and eras.

With the sources of water identified, the veneros, venitas, and manantiales became what are known as the *toma*, where the diversion dam called a presa would be constructed to divert the water to the irrigated parcels of land. The land above the toma would be the commons and the land below would become the irrigated land, and thus private land. The design of the acequia would then follow the contour of the land. Today with some acequias, from a distance it seems that the water is flowing uphill. An acequia usually began at the bend of the river and, as the population grew, kept expanding as another acequia was carved out (at times parallel acequias on both sides of the river), each then becoming a terrace.

Once the toma was identified, an ephemeral presa, or dam, was built to divert the water. As the acequia was being constructed, valves known as desagües were opened on the bank to regulate the water flow. This also allowed workers to use the water itself as a level to see if the water was flowing due to gravity. Sometimes an old bottle was filled halfway with water and used as a level. Once the acequia was constructed, the water could be turned on and allowed to run until it went back into the river through the last desagüe. The first water that ran in the acequia was known as the *puntera*; it was like a tongue wagging along slowly on the bottom of the newly cleaned canal.

The land was given out to settlers in pieces known as suertes that measured two hundred varas wide by four hundred varas in length. A vara is approximately thirty-three inches. (Measurement by feet and inches is a relatively new concept in New Mexico.) A suerte therefore was approximately thirteen acres. These suertes, or fragments thereof, are today known as long lots. Such fragments are also known as *tiras*, meaning "strips," and in other places, such as in the San Luis Valley in southern Colorado, they are known as *extensiones*.

Once the irrigated land, or that land that was below the acequia that could be irrigated, was cleared of brush, piñon, sabina, cedro, *entrañas*, *yuca*, *encino*, or whatever was growing, the property owner would construct a *regadera*, also known as a compuerta, to allow him to divert the water from the acequia madre or acequia secundaria to his particular field. Nothing was wasted; all the brush was used to slow down the water flowing down the arroyos, and over time terraces appeared on the landscape. Remember, nothing was done overnight; this process took time, many years for the land to get to where it is today. It also took a lot of blood, sweat, and literally tears, as sometimes the vecinos lost their life defending their land from other people, or they were bitten by rattlesnakes or even attacked by bears or mountain lions. One has to imagine the landscape then as wild and untamed, with no roads or houses.

Suertes: The Consumers of Water

The acequia has been designed and constructed. The land is clear of all brush and ready to be planted. Now comes the real work, the creation of the terraces. The basic strategy for water conservation in an acequia design is the building or breaking up of the land into terraces, which means that the land has to be leveled so that the moisture is retained by the soil.

From oral histories I have been able to identify four basic types of terraces, some very small and others quite big. Terraces include (1) bancos, usually on slopes; (2) bancales, mostly in valleys; (3) ancones, situated along bends in the river; and (4) *azotellitas*, from the word *azotea*, which means a very small terrace about the size of a flat roof. A bancal can be broken down into altitos, joyas or jollas, vegas, and ciénagas, as we will see later on. Any of these can be further broken down into melgas and then into eras, especially altitos, joyas-jollas, and vegas. Ciénagas can be used for planting if they are drained of the excess water; this is known as *sangrando*. A well-designed landscape can deliver water to the last corner of the property if properly terraced. If the *cabeceras*, *brazos*, *ramos*, carreritas, and surcos are properly maintained, the water circulates starting at the top until it's done its work and the *escurriduras* go on to the river through a desagüe. The best system I have seen is in the Alpujarra in the Sierra Nevada south of Granada in Andalusia, Spain.

"How do you construct a terrace?" I once asked the late Cleofes Vigil as he paused from playing his mandolin, in between relating a *cuento*. Getting serious, he put down his mandolin and said, "Muy fácil [very easy], simply get your shovel and follow the land, the land will be your guide, and throw one shovelful of dirt above you and one below and by the time you get to the end, you'll have your terraza done." It sounded rather easy, and to my surprise it was easy. I was able to carve my place into twelve different levels of terracing. Only recently did I realize I had bancos, bancales, ancones, and azotellitas. The land itself was my best guide and teacher, but I hadn't realized that the answer to my erosion problems was right in front of me until Cleofes made it so clear and simple. He opened my eyes to the memory the land had within. All I had to do was ask the land for guidance and the answer revealed itself, a perfect example of the wisdom of the land.

The summer landscape of the Río Arriba, like that in any desert or oasis landscape, leaves one searching for a *sombra*, plenty of shade, usually in the form of a big crab apple or apricot tree, which takes the place of a *pabellon*, or pavilion. In the summer smell and taste come together without losing their independence, as in a summer *barbacoa* (barbeque) with green chile *rescoldado* (roasted peppers) and corn cooked over the coals to go along with the fresh smell of peaches. The smells unite but don't get confused. For the summer is full of smells, of flowers both planted and wild that squeeze out the sun as they move in the breeze of the afternoon. But the summer can also be full of the smell of the dust of the earth or the wet earth after a summer shower, of the heat and shade, permeated by the fragrant smell of ripe fruit: apricots, apples, and peaches. These wonderful smells seem to descend from the heavens to give us pleasure.

But when the night awakens and the day falls asleep, the freshness of the afternoon snuffs out the drowsiness of the day heat and energizes the soul and spirit. The heavens in summer, whether during the day with its deep blue skies or at night with the stars glittering like diamonds, are as sensual as the spring flowers. The poets talk and write about it, while the artists attempt to capture that elusive light of day. But the table can also express such sentiments, for there are pleasures that are like the day and others that are like night. Summer is like a movie, but one in which the smells come to life; the smells that explode with pleasure and laughter, punctuated by the croak of a frog or the howling of a coyote in the distance while close by the fireflies illuminate the night.

A utilitarian Muslim or Greek garden in many ways resembles a traditional northern New Mexico landscape. If possible the house should be elevated, as Ibn Luyun al-Tuyibi advises, to protect it from flooding when irrigating, in case an acequia breaks or when an arroyo runs during a summer storm. The sad thing today is that people place their houses or mobile homes right in the path of the arroyos, because the arroyos haven't run for several years, and when they flood they always try to blame someone else instead of their own stupidity.

The well and cistern, or aljibe, should also be elevated. Traditionally people would never build on food-producing land or, if there was no other alternative, they would build only on the most marginal space—on a slope or where the soil was very sandy or gravelly. No one would ever think of building on the joya-jolla, the most fertile land, which was set aside for the labor, or *huerta de chile* and milpa de maíz. In the fall the suerte became a rastrojo as the stubble from the corn stalks, chile plants, or whatever else was opened up for the livestock to graze. The livestock not only cleaned the land but also fertilized it with the manure they left behind. This type of landscape always provided a variety of soils, from excellent to not very desirable, as well as access to the river to get water when the acequias were shut off during the winter. People knew what would grow where based on the color and texture of the soil.

In a traditional garden (huerta or milpa) water conservation is critical. In Mesoamerica watering was equally important in the way the waffle gardens, or eras, and chinampas were constructed. Eras were the opposite of the raised beds familiar today. They were sunken beds designed to retain the little moisture of the desert environment. This is a lesson that has been lost on most people, who don't realize that raised beds just don't work in a desert environment.

Terracing has always been an important concept of traditional agriculture, and the Indo-hispano tradition is heir to the terraces of the Alpujarras south of Granada as well as the terraces of Machu Picchu in Peru. Terraces were also abundant around Mexico City during the reign of the Aztecs. Every village in northern New Mexico has terraces, and if they are no longer visible to the untrained eye, once the landscape is peeled back like an onion their outline reappears and they can be reconstructed. Terraces were constructed to hold the soil back and in this way to retain the flor de la tierra, or the most fertile soil, for growing food. Also, terraces were a way of bringing the water in a beneficial way from higher to lower land and of allowing the water to be used again and again, without causing erosion.

The late Cleofes Vigil of San Cristobal mentioned to me that a piece of land had to be plowed at least three times before seeds were put in the ground. The classical agriculturalists, whether Roman or Moorish, recommended turning the soil up to five times before planting.

Another Moorish concept, still practiced by Indo-hispanos, is that of using the garden as at once a flower garden, kitchen garden, and orchard. That type of a landscape, buried under all the brush at my grandfather's property, which is now under my direction, was the genesis of my gardens and what started my research.

Regarding leveling of the earth in a garden bed (*hawd*), the Moorish writer Ibn Luyun in *Tratado de agricultura* notes that the soil should be sloped in such a way as to facilitate the even distribution of water. A rudimentary but efficient device for leveling the bed consists of a cord stretched between two stakes spaced fifteen units apart, with a weighted plumb line (*mizan*) measuring one unit (usually about nineteen inches, which was the measurement for a *codo*—the length from the elbow [codo] to the knuckles) in length dropped from the farthest end of the cord. Luyun writes that bedding plants should be organized in rows, in a rectangular bed, following a one-to-three ratio, reaching no more than twelve cubits in length and less if water is scarce. The Arabs, by using the human body to teach math—besides codos, *dedos* (fingers) and *palmas* (hands) were also used—made sure the farmer had everything needed with him when he worked the land. That is still the philosophy followed today by traditional agriculturalists in the Río Arriba. Here we use *pasos* (approximately three feet), dedos, *manos* (hands), and so on.

Another Spanish Arab, Ibn Bassal, in *Libro de Agricultura*, describes how to plant vines and trees, stating precisely how deep the hole should be, how far apart each young shoot should be, and the dimensions of each garden bed (measured in codos). He also talks about the cultivation of vegetables, melons, legumes, spices, herbs, and flowering and aromatic plants, such as roses, violets, basil, marjoram, and chamomile. He closes with a section on the preservation of fruits.

Bassal, like other Moorish agriculturalists, talks about the importance of *abono*, or compost. Oral history interviews I have done also clarified the importance of allowing weeds to decompose in a hole and then using the decomposed materials for building up the soil. Chinampas were also fertilized by using mud mixed with decomposed plant materials to build up the soil for chile and maíz.

The acequia follows principles similar to those of the human body, in which blood flows from the main arteries to veins and capillaries.

Ibn Bassal, writing in 1075, identified four sources of water: (1) water from rivers that is then channeled into a network of acequias; (2) water from norias (na'ura), or wells; (3) spring water; and (4) rainfall. And the "water way" described by Ibn Luyun is nothing more than an acequia, which is the Arabic word for "water canal." There is nothing more coveted in a property than to have the acequia run close to the house. It has a soothing effect and provides for a very tranquil sleep.

Each acequia can be thought of as a separate terrace. In the Río Embudo within the Embudo land grant there are ten major acequias, or ten major terraces, with the smallest irrigating about ten acres, and the biggest close to two hundred acres. Given that the settlers were very aware of their environment, the toma was chosen for its *venitas* (veins) *de agua*, which pumped water from the springs in the river, or *veta*, the main lode. A presa (dam) diverts the water from the river to the acequia madre. Usually there are two desagües, about a hundred feet apart, to allow for regulation of the water when the river is running high in the summer or during the spring runoff. Some acequias have a third desagüe about a quarter mile from the second, again for an emergency, but it is hardly ever used. If an arroyo happens to run into the acequia, another desagüe is needed to allow the acequia to be cleaned of silt after a flood.

In northern New Mexico the acequia is also used to delineate property boundaries, and such acequias are known as *linderos* or *cequiecitas menores* (to differentiate them from the acequia madres, or mother ditches) when they flow perpendicular and as cabeceras when they flow horizontally along the acequia madre.

When the water gets to the suertes, each parciante has to install a regadera or compuerta (a head gate) to divert the water from the acequia madre to the individual property. Once the water enters the parciante's property, it is spread out via brazos that take the water to the different bancales, also known as ancones (terraces), or melgas, then further broken down to irrigate the eras through small cequiecitas called ramos and finally *hijuelas* or carreritas. Eventually all the water comes together at the desagüe (outlet) that every property has in order to move the water to the next parciante. In the last property there is a desagüe to send the water that is not used back to the river, following the refrán that says, "Agua que no has de beber dejala correr" (Water that you will not drink, let it flow downstream). In some cases one acequia flows into another, as with the Acequia Junta y Ciénaga, which connects to the Acequia de la Nasa, which uses the sobrante, or supposedly excess water.

In the spring, after the limpia, or annual cleanup, is complete, the water is turned into the acequia, a process known as the *suelta del agua*. The water that runs in front picking up all the debris is known as puntera. In the past this was a great, festive occasion in the village, for it meant that the lifeblood of the community was flowing once again. Kids would run ahead of the water yelling, "The water is coming!"

One of the most misused acequia terms is *sangría* (bloodletting; drainage), which a lot of people confuse with a small cequiecita. A sangría is indeed a small ditch, but it is used to drain a ciénaga, or marshland, so that piece of land can be used for cultivation. Like a lot of the concepts pertaining to acequias, this one also derives from references to the human body. When a person smashes a finger or has a swelling that needs to be drained to relieve the pain, one must *sangrar* (drain) the injured area. The same is done with a piece of land that has too much water; it's drained by a sangría. Some people also give the name sangría to a small lateral ditch that runs horizontally to the acequia madre and is used to water a melga. But that is a more contemporary usage.

The acequias menores or acequias secundarias, secondary ditches, that take water to more than one parciante do not fall under the jurisdiction of the *comisión* and mayordomo of the acequia madre.

The person in charge of the water in the Río Arriba bioregion is known as the mayordomo, and he is either appointed by the three *comisionados* (commissioners) or elected by the parciantes, who also elect the comisionados. The mayordomo is under the direction of the comisión. In earlier times he was known as the *cequiero* (sahib al-saqiya, or zabacequiero); he is the one who divides the water, acting like a barmaid and making sure everyone has water. Moorish sources say little about the mayordomo, but Christian documents sometimes refer to the position, as in this reference from thirteenth-century Aragon: "D'aquel qui guarda el agua o la cequia, qui es clamado cavacequia" (The one who takes care of the water who is called a mayordomo).

The mayordomo is always referred to as one who is *digno de confianza* (worthy of being trusted), *el que es fiel* (he who is faithful), or *el fiel del agua* (he who is faithful with the water). In the Hispano-Arabic world, as well as in the Indo-hispano world, water is considered a *don divino* (divine right), which means it belongs to all and ought to be divided equally among those who need it.

Water is always divided based on the amount of land in each acequia, then based on the number of peones each parciante has under that particular acequia and on the amount of water in the river. There are two definitions of *peón*. One refers to a worker, such as one who cleans the acequia in the spring or cuts willows. But the second definition refers to water rights. A peón is a measure that can be divided into quarters. Someone with a quarter peón irrigates one acre or less, one with a half peón irrigates two acres, and so on. Today, with land being divided as subdivisions, the peón formula doesn't always apply.

At times parciantes try to apportion more peones than they have by subdividing the land into more pieces of land than they have water rights. For example, someone with only half a peón (the right to irrigate two acres) might divide a four-acre plot into four one-acre plots; instead, they should divide the land only into two two-acre plots, so that each parcel of land can have one-quarter peón. Dividing water rights into smaller portions than a quarter peón would be a nightmare for the mayordomo and comisión to manage.

The amount of water in a river is measured in surcos. In northern New Mexico a *surco de agua* is the amount of water that can flow through the *buje*, or opening in the center of a cartwheel in a *carreta* (cart). Again, these measures have fallen out of use as land has been divided or sold according to different systems of measurement.

The *repartimiento de agua* is based on the Moorish concept of equidad, which comes from the Qur'an. This concept is based on custom and tradition and is always passed down orally, not in written form. The Tribunal de Agua in Valencia has met every Thursday on the steps of the cathedral of that city for over a thousand years and none of their decisions have ever been written down. In Murcia they have a similar entity called the Consejo de Hombres Buenos, the Council of Good Men. In the Río Embudo we follow a similar practice, though it has no formal name and functions mostly in times of drought. When the repartimiento goes into effect, the water in the river is first divided by the number of acequias (in the Río Embudo there are eight major acequias), then the water in each acequia is divided by the number of acres based on the number of peones held by each parciante. For example, in the Acequia Junta y Ciénaga there are approximately eighty acres under irrigation with eighty quarter peones and at present thirty-seven parciantes (water rights owners), with some having only one-quarter peón (or one share) and others up to two and a half peones (or ten shares).

In years of drought the water used to be divided first by surcos in the river, then by *filas* (or *hilas*, *hilos* here in northern New Mexico) in the individual pieces of land, to irrigate better. Then, in the acequia, these filas (or hilos) are known as *tandas* or *turnos*. The apportionment process during the repartimiento is known as the *tiempo del papelito*, because people receive a paper note telling them when they can have the water and for how long. The papelito is based on the amount of land, which should correspond to the number of peones and acres. In times of extreme drought several hilos of water would be grouped into one (*se jaricaban*) so that there would be enough water to irrigate, as was done in 2002 in the Embudo Valley. The water would be sent down one acequia at a time, with each parciante receiving it for half the normal time, or one hour per peón, but with a full flow.

A *hilo de agua* usually corresponded to one hour of water use. The problem here is that usually the big acequias end up losing irrigation time because the division is not done equally. The water is divided between the upper and lower acequias, with the upper getting it for three days and the lower for four days. Because the number of acres served varies from acequia to acequia, the smaller acequias end up getting more watering time per peón. When several hilos se jaricaban (were combined), they were called *lonjas* (referring to a heavier flow of water; the term literally means a thick and wide piece of the hog's back that has the *chicharron* from which lard is derived). Several hilos were combined into a lonja in order to irrigate better.

In times when there is plenty of water, nobody really cares about measuring it or about how much water an acequia uses, though with water adjudication now a reality, sooner than later water will be quantified. But in times of drought, like in the 1950s and in 2002, acequias have had to fall back on the ancient tradition of adhering to the repartimiento de agua, the sharing of water, based on *la palabra del hombre* (the oral word of man) and equality. When the repartimiento is in force, the comisionados and mayordomos figure how many surcos are in the river at that time and then divide the number of surcos among the different acequias based on the number of peones (which should correspond to acreage) each acequia has. For centuries this system has worked, but during the droughts of 2002 and 2013 some of the newcomers didn't want to follow the custom and tradition. How long the repartimiento system will continue no one knows. Now those with drip irrigation want water every day, instead of following the old tradition of building a holding pond and filling it up when it's their time to water, then using it when needed. Some people also feel that commercial (farmers market) growers should have preference.

The elders, the *viejitos*, talk about *la sabiduría del agua* (the inherent knowledge of the water) and the juicio de la tierra (the wisdom of the earth)—concepts that came through the Moors to the Nuevomexicanos—but that wisdom is rapidly disappearing as the Spanish language, which is the keeper of our environmental ethics and philosophy, is replaced by the English language, and as Roman law, with its underpinnings of Arab custom and tradition, is supplanted by English common law.

Filo de agua, or hilo, like surco, is a concept that is used by the elders, though today the meaning is somewhat different for each person. Huertas de chile are usually watered with a hilo de agua ("thread," a small quantity of water) since you want the water to penetrate deep into the roots and not simply flow to the end of the row before the water is moved to the next row. In terms of water conservation, irrigating with a hilo de agua is equivalent to drip irrigation promoted by proponents of sustainable agriculture. The former requires wisdom of the land, coupled with the knowledge of the water, while the latter requires big investments of capital. So which is really more sustainable? The concept of hilo de agua is an Islamic concept that some scholars say comes directly from the Qur'an. But my father also adhered to that concept when irrigating his huerta de chile. Since water is heavy, if a lot of water is used to irrigate, the soil will get compacted, and if watered too often, the plants will turn yellow and not grow. One has to wait to water until the plants *piden agua* (ask for water), which is when the leaves start to wilt. And gardens should always be watered very early in the morning or in late afternoon, but never during the heat of the day. This is something recommended by the Arab agriculturalists, Gabriel Alonso de Herrera in his classic work, and the viejitos of northern New Mexico. The only time that a garden is watered during the heat of the day is during the repartimiento de agua, when a parciante is given a papelito (which can be at noon or at midnight). But that's because in northern New Mexico the concept of using an alberca, or pool, to collect the water to use at more appropriate times has all but disappeared. Yet there are examples of people using albercas to store water.

Two other very important concepts in terms of the philosophy of the sharing of water are sobrante (excess water) and *auxilio* (sharing, or coming to the rescue of those who don't have enough water). Usually when a new piece of land was exploited, it was watered with the sobrante from an already established acequia. Again in the Embudo Valley, the farmland in La Nasa is watered with the sobrante from the Acequia Junta y Ciénaga. But the sobrante can also be applied in times of drought, when there might be more than enough water for one or two acequias that might have the water for that particular turn, or turno. Once two acequias are *surtidas* (full to capacity), the excess water is known as the sobrante and the next acequia in line can capitalize on this water. But once it enters the acequia then it is up to the mayordomo to follow the established turnos in terms of irrigating with that sobrante.

Auxilio comes into play when a certain acequia doesn't have water and the gardens are drying up; then the comisión from that particular acequia can petition the comisión of the acequia that has water for an auxilio (usually a one-time help), to let them have some water to save their gardens. The water that is shared in times of need (auxilio) is not sobrante, or excess water, though New Mexico law does not recognize sobrante, since the water is already over appropriated. If an acequia has plenty of water, instead of sending the water into the river at the end of that acequia, the water is allowed to run into another acequia, but first rights belong to the upper acequia. Also, as in the case of the Acequia Junta y Ciénaga and the Acequia de la Nasa, both are independent of each other, with separate comisiones and mayordomos. Acequias, which appear easy to understand, in essence are very complex to comprehend and manage.

El Altito

First-time visitors to northern New Mexico are intrigued by the division of the landscape, especially the long lots that are so much a part of the historic vernacular landscape, as are the adobe houses and hornos. The history of the long lots goes back to the land patterns of Spain and even further to the way the Arabs saw the land, which comes as a surprise to most people, even the native born. There are four basic components to a suerte, though not all suertes have all of them. First, immediately below the acequia is the altito, or highland, used mostly for orchard crops, if trees are planted in rows. But trees are also used as windbreaks and to separate one terrace from another. The altito, then, consists of the first terrace. The other three components are the joya-jolla, vega, and ciénaga.

El Huerto/The Orchard: A Middle Eastern Inheritance

William W. Dunmire, in his 2004 book *Gardens of New Spain: How Mediterranean Plants and Foods Changed America*, writes that apricots, cherries, nectarines, and peaches were grown in New Mexico as early as 1630, according to the Portuguese Franciscan Fray Alonso de Benavides, and plums were grown in San Gabriel prior to 1600, according to Villagrá. Apples seem to have been growing in Manzano (which means apple) as early as 1633. The Arabs, along with quince and pear, introduced apples into Spain. Pears are not mentioned in New Mexico prior to 1776. Marc Simmons mentions quince during the colonial period but gives no specific date. Grapes were grown in the Socorro region as early as the 1620s, according to Simmons.

Supposedly an elderly man told the "Fig Man," Lloyd Kreitzer from Albuquerque, that the Spanish settlers brought several types of trees with them, along with a shrub that is found throughout northern New Mexico. The trees or vines brought by the settlers were the vid (grape plant for wine), fig, pomegranate, and plum. The nonfruit shrub was the lilac, which the settlers used as a barometer to know when to plant. In the spring when the lilac starts to leaf, that means it's time to start planting the spring garden, and when the flowers dry on the plant and start to fall, it's time to start the fall garden. There are about six weeks between the time lilacs start leafing and the time the flowers start falling off. The importance of the lilac is that it is site-specific, that is, it will leaf and flower at different times depending on the specific microclimate where it is planted, based on the elevation and whether the plant is facing south (resolana) or north (sombrillo).

Land documents from northern New Mexico refer to three different types of soil on the agricultural land. First is tierras de pan cojer, which usually refers to lands that can produce two crops per year. Here in this part of New Mexico a joya-jolla might fall under this category. Second is tierras de pan sembrar, or other irrigated lands. Third is tierras de pan llevar, or those lands that are dry farmed or used for grazing.

One of the most neglected aspects of northern New Mexico history is the influence of the Arab world, which is etched not only in the language and landscape but also in the fruits and vegetables that grow in this semiarid land, given life by the flow of water from the acequias. When people think about northern New Mexico, most think of the Spanish influence without realizing that the Castilian nation was born in the same year that Columbus made his famous journey across the Atlantic. Up to 1492, Spain was a multicultural and multiethnic conglomeration of Christians, Muslims, and Sephardic Jews, and how the land was viewed, organized, and worked was influenced more by the Arabs than the Castilians.

The first time I became conscious of the land-division patterns in northern New Mexico and how different they are from the rest of the country was in the 1950s. My father was reading a Nancy and Sluggo cartoon. Nancy had just bought a piece of land one mile long and one inch wide, and Sluggo asked her, "What are you going to plant in such a piece of land, so narrow and so long?" To which she replied, "Macaronis." My dad broke out into a hysterical laugh and said, "Apparently Nancy bought a piece of land in *el norte*, because that's the way the land is here."

The land grants are divided into common lands and private lands. The irrigated parcels known as suertes originally comprised the private lands, and these divisions have their origins in the Middle East. To the Arabs who settled in the Iberian Peninsula the irrigated or appropriated lands were known as *mamluka*; in the New World they became known as suertes (luck) because they were given to the settlers of the land grants based on a lottery system. Under Spanish law the suertes were those lands that fell below the acequias, and they were therefore the irrigated pieces of land. (Mexican grants seemed to have operated differently, as with the Sangre de Cristo land grant in southern Colorado, for example.) It was the acequia, then, that divided the private lands below from the common lands above.

The suertes were set up as long lots so that everyone could have access to the river and to the commons; this type of land distribution made sure everyone had good land for growing crops but also land for domestic animals such as a milk cow and a few sheep to graze close to home. Suertes were then divided into the altitos, or highlands where fruit trees were planted, and the joya-jollas in the lower lands. Still today you find people referring to the best land on their individual parcel as "la joyita."

Velarde in the Española Valley was originally known as La Joya due to its fertile lands. There is also a Joya by Belen. Below the joya-jolla was the vega, which can be used for planting but in New Mexico is most commonly used for pasture for domestic animals. Below the vega was the ciénaga, or the marshland. Ciénagas can also be used for growing crops if they are drained, or *sangradas*. Then came the esteros and bosques, right at the river's edge.

This type of land division was not oriented toward growing for a market but rather toward providing for the community, which was usually a very tightly knit society based on familial ties. In a way it was an intentional community, like those that are now the rave among the rich in the Santa Fe area, but these intentional communities were composed of campesinos, or rural people, farmers whose pieces of land were rather modest, comprising only a few acres. These land holdings were known as minifundias, in contrast to the landholdings of northern Spain known as latifundias, which were very extensive and usually used for livestock grazing. In Spain a minifundia under the care of the Moors could feed a family of four, but when taken over by the Castillians, who were unfamiliar with this type of irrigated farming, it could barely feed one person.

As landholdings of this type are sold to people outside the family, conflicts arise, because the newcomers have a totally different cultural background and orientation. As a result the suertes are disintegrating as the traditional land divisions of the long lots give way to horizontal divisions in which the altitos, joyas-jollas, vegas, and ciénagas are sold separately. Those pieces of land are becoming landlocked and destroyed by housing and roads.

The traditional land divisions served a purpose. Each type of land offered a different microclimate, and traditional landowners knew exactly what could or could not be grown on each division. Fruit trees, for example, were usually planted on the higher land of altitos, where they were less susceptible to freezes, as the cold would settle on lower lands toward the river. Another important factor was that the irrigated land was never used for housing. Now, since a lot of the land grants are no longer intact and the Bureau of Land Management and Forest Service now own most of the common lands, and because the population has grown, the cultivated lands are becoming residential suburban-style lots.

Vega, then, refers to lowland that is humid and level, or llano; the name comes from the word *vigore*, because it's vigorous and fertile. In Arab it signifies *a tierra de labor puesta en llano*, or a level land that is worked, or planted. In New Mexico it refers more to an irrigated pasture, whereas in Andalusia it's where food is grown, such in the Vegas of Granada, Valencia, and Murcia. And ciénaga comes from the word *cieno*, which is usually a black mud, smelly and soft, which is neither mud nor water and without draining can only be used marginally for grazing.

Ideally most suertes would be composed of altitos, joyas-jollas, vegas, and ciénagas but not all contain all four types of land, especially nowadays, as the land gets broken up into smaller and smaller parcels. But they also contained bosques and sometimes esteros. Ciénagas, bosques, and esteros (estuaries, from the Latin *aestuarĭum*) acted as purifiers, cleaning the water before it went back to downstream users.

Also, there are different types of terraces or terrazas, also known as bancos or bancales and ancones. Bancos and bancales run along the valleys and on slopes, and those by the meandering of a river are known as ancones. Terraces are irrigated by diverting water from the acequias; therefore, it is the acequias, because of their rigid design patterns, that give birth to the suertes.

Historically most suertes, as mentioned earlier, extended from the acequia to the river, but there are also places in the Río Arriba bioregion, such in the San Luis Valley of southern Colorado, where the suertes extend above the acequia. In San Luis these long lots are known as extensiones, or extensions.

This type of agroecosystem, whose roots can be traced to the Fertile Crescent, with modifications made in southern Spain and then in Mexico before it finally arrived in New Mexico in 1598, is now on the verge of disappearing. Very few people now know how this type of system operates, and fewer still know the history of its origins.

A similar system of land division also exists in Hawaii, known as *ahupua'a*. It is more in tune with the extensiones of the San Luis Valley, since the strips of land in Hawaii originate in the mountains and go all the way to the sea.

The *Gioias* or Joyas-Jollas: *El Jardin*/The Garden

Now let's further dissect the suerte by focusing exclusively on the joyas-jollas. The name comes from the Tuscan word *gioia*, which means happiness but also something very precious, such as a jewel (joya). But this "jewel" is also a jolla, a hollow, a small valley between mountains. Thus the way it's interpreted in New Mexico, this type of landscape is a very fertile jewel. Translated to land this meant the most fertile lands, where people usually plant their chile and other vegetables for home use or in the past to trade for what they didn't have. This form of barter, still practiced today, is known as cambalache, from the word *cambiar*, to trade.

Traditionally the joya-jolla lands were used for growing food; the huertas (large vegetable gardens) and *jardines* (small gardens) were planted in these strips of irrigated land, also known as *tablas*. The term *huerta* comes from the Latin, deriving from the term *hortelano*, meaning a person who works the land for food. Jardin, or flower garden, from *riardin*, is more closely tied to the Arab concept of a garden. Some say the name derives from the German, which was adopted by the French as "jardin." And in northern New Mexico we added the concept of milpa, from Mesoamerica, referring to a cornfield.

Thus the joya-jolla was where the huerta de chile and milpa de maíz were planted and also the *melonar* (melons and watermelons), which had to be planted in sandy soil. The farmer knew his land like the palm of his hand; he knew where he could plant what, because he understood the microclimate and soils of his place.

The joya-jolla was further broken down into melgas, from *mielga*, a word that came from Italy, from the region of Media, a corruption of the terms *medica herba*, which was a common pasture plant for animals. The Arabs called the plant *alfalfasat*, or alfalfa. In New Mexico the melgas were strips of land broken down into manageable parcels, usually fifty feet in length and the width of the suerte, where alfalfa was planted. An *emelga* is the land between two *sulcos* or surcos, the land between two furrows. When it is part of the joya-jolla, a melga can further be broken down into eras as a water-conservation strategy. There are two

Milpa. Photograph by the author.

types of eras. One, which is located in the commons, is for threshing wheat or other grains. The other is in the form of a sunken bed. Among the Zunis such beds are known as waffle gardens.

An era usually refers to the place where the hortelano plants lettuce, radishes, and other vegetables, and it is also known as an Afghan garden, which looks like a comb. Today these types of beds are still found in New Mexico and also in the outskirts of the city of Chihuahua, Mexico.

Milpa, the Mesoamerican Influence

From the Americas we got chile, maíz, frijoles, calabazas, papas, *tomates*—foods that are among the daily staples of northern New Mexican cuisine. Traditional Indo-hispano cuisine in northern New Mexico is complex, built from several different layers of influence, starting with the Roman past, then the Moorish and Sephardic influence and our own

Mesoamerican traditions, until what emerges is a *platillo* known as *comida nuevomexicana*. Though what one gets in New Mexican restaurants is called "Spanish food," it has no resemblance to the food one finds in Spanish restaurants. Also, why restaurants got into the habit of piling cheddar cheese on New Mexican cuisine, I don't know. Traditional New Mexican food prepared at home is void of yellow cheese; if any cheese was used, it was the homemade *queso blanco*, or white cheese.

Potatoes, which are a staple in most Hispano households today, came relatively late to New Mexico; they are not mentioned until 1831, by Josiah Gregg. Similarly, tomatoes are not mentioned until 1745, when they were growing in Santa Fe. Both crops also did not take a hold in Europe for a long time, as people thought both of them were poisonous.

What's *agricultura mixta tradicional mestiza*? In terms of "traditional" Indo-hispano agriculture, it starts with the land grants, which are then broken down into the commons, acequias, and suertes. But to be more precise we have to understand the huertas (fruit and vegetable fields), jardines (gardens), and milpas (corn fields), which is where we find the Roman, Moorish, and Mesoamerican traditions all combined to form a hybrid traditional agriculture that is northern New Mexico agriculture to the core. I call this form of agriculture "agricultura mixta tradicional mestiza."

It's "mixta" because it combines fruit trees, vegetables, and legumes, along with livestock, fowl, and bees. This mixed type of agriculture developed in New Mexico around the land grants made by Spain and later Mexico, which included both irrigated and dry farming, not only growing fruits and vegetables but also grazing animals such as Churro sheep, Corriente cattle, pigs, and turkeys, native to the Americas. Of course, it also included foraging for wild plants, hunting, and fishing.

It's "tradicional" because its roots are organic, sustainable instead of subsistence (as our agriculture has been called), because it has sustained Indo-hispanos for centuries. Since it adheres to the old methods it follows the precepts of permaculture. Here the acequias and terraces anchor the permanent agriculture of this system.

It's "mestiza" because it's a hybrid of Old World and New World systems (acequias and chinampas), techniques (surco and *tapanco*), fruits (cherries and capulín), vegetables (lettuce and tomatoes), and animals (chickens and turkeys). New Mexico's agricultural tradition is a blend of the agricultural practices of Andalusia (based on Asian, Latino, and Arabic models) and Mesoamerica (Tlaxcalteca-Pueblo). Therefore, the concept "mestiza" combines fruits, vegetables, and methods from both sides of the Atlantic. It is Roman, in that the term *huerta* is a Latin concept, *hortus, ti,* from the verb *orior,* "about to be born"; that is where the vegetables and fruits are born and raised. Milpa is a Mesoamerican idea, the place where corn is planted (from the Náhuatl *milli,* a seed bed, and *pa,* "where"; in this case a "place where corn is sown"). Chinampas are still used to grow corn; they are nothing more than big beds constructed in the middle of a lake, made out of mud dug up and piled into beds. Thus they work like hydroponics in the way water is absorbed. If the chinampas are too low they get flooded, and if they are too high, the plants won't get enough water. The method of watering a huerta, a Roman concept, is by using an acequia, which is Moorish. An *almácigo* (a place to start plants early, usually inside the house, as my mother did) is also part of the agricultural vocabulary in the Río Arriba, and it is Arab in origin.

A perfect example of our Mestizo heritage is a sentence a father might say to his son without thinking about the etymology of words: "Agarra la pala y haz un tapanco en la cequiecita" (Take the shovel and make a heap of dirt to divert the water to the small ditch). Breaking down the sentence, *pala,* "shovel," is a Latin word whose roots are Hebrew; *tapanco* comes from the Náhuatl *tlapantli,* "a heap or pile of dirt"; and *cequiecita* is an Arabic word that means a small canal to transport water. That simple sentence uses words in Castilian, Latin from the Hebrew, Náhuatl, and Arabic.

With the coming of the settlers under the Spanish Crown, a group that included Tlaxcalteca Indians and people of other European nationalities plus mestizos from Nueva España, came a lot of different fruits and vegetables that had never been seen in this part of the world. When Oñate settled in Ohkay Owingeh in 1598, which he renamed San Juan de los Caballeros, the settlers carried with them their favorite seeds, plus they also brought with them sheep, goats, cattle, pigs, and horses, revolutionizing the agriculture of the area. Radical historians and scholars have vilified Oñate, and rightly so, but they don't give him credit for the agricultural revolution he introduced.

Along with the different plant seeds and trees, the settlers also introduced different irrigation techniques and work instruments such as the plow, forever changing the agriculture and diets of the Native Americans. Before the arrival of these European and Mesoamerican settlers, the main food staples of the Native population in the Southwest were the trilogy of corn, beans, and squash, known as "the three sisters." Of course, they also hunted game and foraged for wild plants and fruits. As is widely documented, within a month of settling in Ohkay Owingeh Oñate had 1,500 indigenous people and new settlers digging what is believed to be the oldest acequia in New Mexico, on the Chama River in what is now the community of Chamita.

Oñate knew that in order for the new settlement to survive and thrive, they needed water. According to Dr. Tomás Martinez Saldaña, an agricultural historian from Mexico City, the Tlaxcaltecas were instrumental in helping lay out the acequia systems in what is now New Mexico.

The encounter with America resulted in an enormous exchange of species between the Old World and the New World. It is estimated that 40 percent of modern economically relevant crops originated in America, a fact that sometimes makes it difficult to imagine Old World culture and gastronomy without American-originated crops. For example, corn, sunflowers, potatoes, tobacco, peanuts, cocoa, beans, squash, pumpkins and gourds, tomatoes, chile, and many other crops were new to the Old World a few centuries ago. Many Old World crops—such as soybeans and coffee from Africa and Arabia and bananas from Southeastern Asia, not to mention oranges, limes, sugarcane, and salad greens—also adapted well in America, and this continent has become the main producing area for them. Christians talk a lot of the Last Supper, but what about the First Supper? That is the first time those under the Spanish Crown ate a combination of foods from both sides of the pond. José Luis Monteagudo wrote a book dealing with the concept, *La Mesa de Hernan Cortes*.

When the Moors were driven out of Iberia, Castilian agriculture went into a tailspin. The old Christians had no idea how to grow food, since they had been used to mostly dryland farming, or secano, instead of irrigating to produce food. The Tlaxcaltecas were also exceptional farmers in their own right, but they quickly adopted the new techniques, vegetables, and fruit trees to their environment.

Just as the Moors had introduced the technique of grafting to the Iberian Peninsula, the Spanish settlers introduced it to the Americas. Here in New Mexico people started grafting the new varieties of apples, peaches, and apricots by using *trementina*, sap from the piñon trees, for grafting purposes. They also used, and I still use, mud from black earth when a fruit tree is damaged in order to heal the wound, and it works wonders.

Manzanas, apples, were first introduced to the Americas by the early settlers; they had been brought to Spain in the tenth century by the Arabs, who knew them as *tuffah*. Spanish settlers in the Americas also introduced the *albaricoque*, or apricot, along with the peach, known as *durazno* and *melocotón*. Watermelons (*sandias*), and melons (melones) were also brought to the Americas from across the ocean. In fact, when the pobladores first came to New Mexico they found the Native Americans already growing peaches, melons, and watermelons. These fruits had preceded the settlers, as the indigenous people had traded for the seeds in Central Mexico.

La Huerta de Chile

The acequia landscape in New Mexico cannot be described or understood without considering the huerta. In the Española Valley, the heart of Pueblo country and the place where the first European settlement was established in 1598, huerta and chile are synonymous. Every New Mexican landscape is defined by a huerta de chile and a milpa (cornfield), often planted side by side, and at times only the farmer can tell the difference. Some farmers, without historical knowledge, plant a row of corn, then two or three of chile, another row of corn, and so on, following the model of Mesoamericans in their chinampas at Xochimilco.

In the Iberian Peninsula a huerta usually refers to irrigated agriculture, more than likely in vegas, like the huertas in the vegas of Valencia, Murcia, or Granada. In northern New Mexico a huerta can be defined as a plot of land—usually in the joya-jolla, or most fertile land—that grows chile along with tomatoes, cucumbers, and other vegetables. You don't even have to mention chile, just say huerta, and people familiar with New Mexican traditional agriculture will immediately conjure up a chile patch. When *huerta* came to mean a chile field I haven't been able to find out, since in Mesoamerica chile was part of the milpa. According to Covarrubias, a huerta, or *güerta*, is a piece of land that has water for irrigation, while a piece of land with flowers is known as a jardin, or garden.

Sadly, most of the terminology related to the huerta de chile has gone by the wayside as the New Mexican Spanish language has disappeared and new technologies have been employed by the farmers. The concept of *matear*, the planting of the chile seeds directly in the soil, has now been replaced, as seedlings are now grown in greenhouses. As a result the calendar followed by the hortelano has also been forgotten. The greenhouse has also replaced the almácigo, or the bed where seedlings were grown.

Besides corn, or maybe even more than corn, chile is the most important crop to native New Mexicans, just as it was to the ancient Mexicans. In the cities of Teotihuacán, Tula, and Monte Albán, evidence of ample consumption of chile has been unearthed. The Aztecs had developed a culture of chile, some aspects of which are still visible today. Fray Bernardino de Sahagún, in his monumental *Historia general de las cosas de Nueva España*, written during the second part of the sixteenth century, explains that chile was not only an important part of the Aztecs' diet but it also had a variety of other uses. Members of the military would throw chile into the fire during interrogations, where its smoke served the same purposes as do gases today. Chile also had medicinal uses, commercial value, and even served to straighten out kids who would not mind their parents. In the valley of Mexico chile was sown primarily in the chinampas, and in New Mexico, though not sown in chinampas, the traditional huertas followed the same pattern.

The goddess of chile was called "respetable señora del chilito rojo" (the respectable lady of red chile), and she was the sister of Tláloc, the god of rain, and Chicomecóatl, goddess of *mantenimientos*. Chile also had to do with Tlazoltéotl, the Aztec goddess of carnal love. Sahagún himself notes that during the festivities of Macuilxóchitl, the god of flowers, dance, and love games, men and women abstained from eating chile during four days of rigorous fasting. The ardent god would punish those who broke the fast by making them ill "in their private parts." Some Native tribes still observe fasting rituals that prohibit the eating of chile.

Native Americans knew a wide variety of ways to prepare chile. Generally they prepared it in the form of salsas (*moli* in Nahuatl), in a wide variety of colors, smells, and textures. They had *chilmollis*, salsas of smoked chile, hot chile, dry chile, medium chile, green chile, yellow chile, red chile, and black chile. They also consumed it as a chile beverage, such as *chileatole*, or a combination of fine chocolate and chile water, called *chicacalhuati*. They also had a variety of classifications for the heat of the salsas, from *salsas picantes* (hot), through *muy muy picantes* (very very hot), and *ardientemente picantes* (burning hot), to *picantísimas* (extremely hot).

Here in the Americas the settlers came upon totally new fruits and vegetables. Among the most notable was chile, with the first seeds making their way north with Oñate's settlers and even earlier with Obregon, who brought chile seeds with him in 1580. In Mexico the Europeans were also introduced to the tomato and eventually the potato from the Andes region. In the Americas the Mediterranean trilogy of wheat, grapes, and olives met the Native American trilogy of corn, squash, and beans and since then these have been the main ingredients of most New Mexican food, with the exception of olive oil, which was replaced by lard from the pigs, which also introduced by the Spanish settlers.

Herando Alvardo Tezozómoc, in his *Crónica mexicana*, written around 1598, says the Mexicas planted for the first time in chinampas in Texquiquiac, in the northern part of the valley, in the thirteenth century. Not much later they planted chile in Xaltocan; they planted along the edges of the chinampas where they also grew corn, beans, and squash. The chronicles of the time emphasizes the quantity and diversity of the chiles that were grown in the pre-Hispanic markets. Chile was also used as a tribute prior to the arrival of the Spanish. This was a custom imposed by the Aztecs on those they conquered. Each locale contributed what they grew, most commonly corn, beans, and chile, according to Dr. Janet Long-Solís, an anthropologist with the National Autonomous University of Mexico who has written two books on the history and cultural use of chile in Central America, *Capsicum y cultura: La historia del chilli* and *El placer del chile*. At the fourteenth biennial National Pepper Conference in San Antonio, Texas, in 1998 she spoke of the use of chile going back thousands of years, as evidenced by Mayan and other ancient rock carvings.

"Burning piles of chile were used to discipline children and to fight off enemies. Chile was used to pay taxes in addition to its use as food," she said. In addition, she talked about the rain god of the Aztecs, who is typically depicted holding stalks of corn and chile, and how offerings are made to San Francisco, the patron saint of chile farmers, at the current Fiesta of the Chiles that is held in October in the mountain villages of Guerrero state. "The chile ceremonies are part of the life cycle from birth to death," she said. "Chile is also used to cure several illnesses and to purify a body before burial. But with the modernization of many areas of Mexico, these ceremonies are becoming fewer and may soon be lost," she noted sadly.

After the conquest the Spanish adopted and introduced the same system of tributes, though they used European or Arabic systems of measurement, such as fanegas, *celemines* or *almudes*, *venguenes*, and *arrobas*.

They collected tribute in chile, although over time monetary tributes were introduced.

Today most of the germ plasm introduced by the early settlers has been lost, but there are still some places along the Camino Real where old seed varieties and heritage fruit trees survive. According to Dr. Martinez Saldaña, northern New Mexico is the only place in New Mexico where some of the old germ plasm still survives. Other places along the Camino Real where some of the old fruit, vegetable, and animal species as well as traditional agricultural techniques and acequia irrigation traditions still persist include Valle Allende on the Río Florido in southern Chihuahua and in Bustamante, by Monterrey in the state of Nuevo León. Oñate and his settlers spent a couple of years in the area of Valle Allende starting in 1596, thus literally planting the same seeds and fruit trees that eventually made their way to the Río Arriba region.

The people of the Río Arriba region have been very inventive in times of scarcity, as evidenced by oral history. When there was no coffee to be had, people would roast garbanzo beans and substitute ground garbanzo powder for coffee. They would do the same when chocolate was hard to come by; they would also roast barley, grind it, and drink it as if it were chocolate. Of course there was always ground blue cornmeal from which the settlers made atole and chaquegüe, as is still done today.

Among the crops that Villagrá mentioned as growing in San Gabriel, across the river from Ohkay Owingeh, in the early 1600s were barley, artichoke, cabbage, lettuce, carrot, garlic, onion, radish, turnip, cucumber, garbanzo, pea, chile, and cumin. By 1625, when Benavides came through New Mexico, the settlers were already growing haba or fava, lentils, and vetch to go along with what Oñate brought with him.

The earliest mentioned legume in New Mexico is wheat, which was already growing in San Gabriel in 1599, a year after Oñate settled there, and therefore appears to be the first crop to be grown by the Spanish settlers. Apparently the other crops were not planted until the following year, which makes sense, since the settlers had to make sure the acequias were running and the fields were ready for the exotic vegetables they had brought. Just as they had done eight hundred years earlier when the Moors brought them to the Andalusian region of Spain, these vegetables had to acclimatize to a new locale. Though there is no mention of almunyahs, which were private, experimental, and recreational gardens introduced by the Moors to acclimatize the new fruits and vegetables into the Iberian Peninsula, the almácigos served that purpose and still do.

With the introduction of new crops such as grains, leaf vegetables, root vegetables, "fruit" vegetables, legumes, fruits (including stone and and other orchard fruits), and culinary and medicinal herbs and spices, the diet of the people in New Mexico changed. Even today foods that are known as Indian foods, such as the mutton stew made famous by the Navajos, would never have been possible without the introduction of Churro sheep, along with cabbage, carrots, onions, and other ingredients whose roots are in the Mediterranean.

Arroz con leche, or rice pudding, and capirotada, bread pudding, have their roots in the Middle East, as do buñuelos and sopaipillas, now coming under attack as Indian fry bread. But the main staples of New Mexico cuisine still remain American in nature: chile, whether green or red, fresh, roasted, dried, or frozen; corn, fresh, as sweet corn on the cob, roasted, made into a flour to make atole or chaquegüe, or horno cooked and made into chicos or dried and turned into posole; and of course potatoes are used every which way to compliment the other basic foods. Tortillas, made from wheat or corn, are the bread of the Americas, though tortillas Spanish-style, as omelets, are also part of the diet.

Then, of course, people added beef with the introduction of the Corriente cattle, while pork meat became a main ingredient in tamales, along with chicharrones and lonjas that are used in cooking posole. The blood of pigs was used for making *morcilla*, or blood sausage. Also added to the Native American meat diet of buffalo, antelope, deer, and turkey were chickens, which provided eggs in addition to meat.

Since olives couldn't be grown in the Río Arriba bioregion, the hog became their substitute, providing lard for frying all types of foods. But now the hog and its lard have been demonized. "Even in Mexican and Latin American bakeries with Spanish-spoken-only signs, where the bakers surely know that in their native countries the most savory *empanadas* and the airiest tamales rely on lard, my hopes are usually dashed," according to an August 12, 2005, *New York Times* article by Corby Kummer, headlined "High on the Hog." "Besides," she adds,

> lard seems old-fashioned—redolent of poverty and its companion cuisines. . . . I have a suggestion for those Old World cooks who are wrestling with New World advice: take another look at the fat profile of lard. It has half the level of saturated fat of palm kernel oil (about 80 percent saturated fat) or coconut oil (about 85 percent) and its approximately 40 percent saturated fat is lower than butter's nearly 60 percent. As with all dietary advice, the fat of the day will change. But eternal truths will remain: food is always best with little or no processing and eaten as close as possible to where it is grown. This goes for lard, too. The artisan pig farmers whose fortunes have been revived by a new market for pork with real flavor should look into selling lard because the supermarket kind is processed and dismal. Chefs and short-order cooks can do everyone a favor—even the guardians of the public health—by reaching for the fat that everyone knows tastes the best: lard.

I know because we had a restaurant and people would ask, "Do you use lard?" People asking were usually Seventh-Day Adventists, those who fancied themselves as organic connoisseurs, or vegans who don't want to taste any animal byproduct. On the other hand the local nuevomexicanos would say, in a hushed tone of voice so as not to be heard, "The food with lard tastes a lot better, but they say we are not supposed to eat it, that it's bad for your health."

Old varieties of Mission grapes and wheat are still grown throughout the Río Arriba and are coming back to life as heirloom varieties. When the wheat industry was revived in Costilla in the late nineties, over eighty varieties of local wheat were identified, although the growers went for Kansas wheat so the local varieties all went back to shelf. By now they are probably all lost.

Today's New Mexican cuisine is called a fusion cuisine, as it blends the best from the Old World with the best of the New World, limited only by the imagination of the cook. The rave about organic produce is nothing new to New Mexico farmers, as traditional agriculture has always been organic, though it never used the name. What is grown in a traditional Indo-hispano garden also reflects the influences from many continents: wheat (which was grown extensively in northern New Mexico up to the Depression) originated in Egypt; melons and watermelons came from Africa and Asia; lettuces and other greens from the Mediterranean basin; and brassicas originated in northern Europe and for a long time were the main vegetable in a Christian garden.

Natural Farming: *Jardín Ricio*/Volunteer Garden

There are several marginal crops of Mesoamerican and Mediterranean origin that over the centuries have acclimatized themselves to the arid southwest, especially the high desert of northern New Mexico that are at risk of disappearing. Maybe these crops will not disappear from the landscape, but definitely they are disappearing from the table. Today they are more of a delicacy, with the exception of one that is available in all supermarkets, and mostly prepared only by traditional families who still know what the plant looks like and how to prepare them. I am referring to the *quelite pardo* (wild quinoa), *quelite del burro* or *quintonil* (wild amaranth), verdolagas (purslane), and *esparragos* (asparagus) that grow wild in orchards and in chile huertas and milpas. These four crops are what my mother would refer to as *dándose rizas*, or plants growing naturally with no one having to plant or tend to them. From *ricio*, Latin *recidīvus*, or *renaciente*—springing anew. All that is needed is a good rainfall and all of a sudden what had appeared to be a barren landscape sprouts overnight with all types of edible plants. With nice summer rains verdolagas sprout everywhere during the summer. When I was growing up I remember my mother harvesting armloads of verdolagas, quelites, and esparragos, though the asparagus usually grew on the banks of the secondary acequias or linderos and in the orchard under the apple trees. But with the advent of the tractor and pesticides a lot of the asparagus disappeared.

Now, as more farmers are turning to drip irrigation and farming is no longer a cultural activity but a business venture, the delicate flavor of quelites pardos, a relative of the quinoa plant also known as lamb's quarters, can be found only in traditional huertas that still irrigate by surcos, or in furrows. These plants don't thrive in a drip-irrigation system, as such systems deliver water only to the plants that can be taken to market. Drip irrigation is killing a lot of the biodiversity common to the northern New Mexico upland-desert environment. And with it diets will also change.

Espinacas (spinach) are not the same as quelite, from the Nahuatl word *quilitl*, which is a generic name for edible plants. There is even a song about the quelite:

¡Qué bonito es el quelite!	How beautiful is the quelite plant
Bien haya quien lo sembró	Appreciative of whoever planted it
Que en sus orillitas tiene	That along its edges
De quien acordarme yo	It reminds me of whom to remember

There is also a saying among the campesinos about the quelite, "Quelites y calabacitas, en las primeras agüitas," meaning these plants start producing with the first spring rains. When welfare first came to New Mexico, for some reason the check was late; in late spring people would say, "No importa, ya viene, Mr. Kelly," meaning that it didn't matter much because the quelites were in abundance. And quelites New Mexico style with a few red chile pods crushed with seeds for texture and color and finely minced onions with a plate of beans and freshly made tortillas is still a delicacy that can't be found even in a five-star restaurant. In New Mexico we eat two types of quelites, also consumed in Mexico: one is a wild amaranth that we call *quelite juz* or quelite del burro, known in Mexico as quintonil. Then there is quelite pardo, which is the one consumed the most; in Mexico it's called *quelite cenizo.*

Other plants from the jardin rizo, the original farmers market, are the herbs oregano, *berro* (watercress), chimajá (wild parsley), altamisa (tansy), manzanilla (chamomile), cilantro (coriander), malva (mallow), and *epazote* (Náhuatl, *epatl*; this herb is used mostly when cooking beans).

CABAÑUELAS: JEWISH AND MAYAN METEOROLOGY TRADITIONS COMBINED

Before there was TV or radio, people used to rely on the stars, sun, and other natural phenomena to better understand when to plant or how much seed to commit to the soil. Part of that knowledge is embedded in the system known to Spanish-speaking people as *leyendo* (reading) *las cabañuelas*. The cabañuelas are a system of predicting the weather for the coming year based on observation. They are nothing scientific, but usually very accurate. In fact, the cabañuelas might hold the key to addressing climate change if we learn how to read them or decipher them in a new context.

For the early settlers who braved the Jornada del Muerto, or Journey of the Dead, from Mexico to northern New Mexico in 1598, it wasn't simply a matter of finding water and good soil in which to plant in order to survive; it also meant learning to understand nature in the new environs and applying the knowledge they brought with them to a new site. After all the sun, the moon, and the stars were still the same and the arid landscape was similar to that of Mexico, Spain, and the Middle East.

This knowledge of place, or querencia, ran through their blood. They also had an empirical knowledge of the weather, a knowledge that is basically oral. "January has the secret of all twelve months," or so says an old proverb in Spanish. Or it could be that the month is August, depending on what part of the Spanish-speaking world one lives in. But the dicho is referring to the reading of the cabañuelas to predict the weather for the coming year.

The settlers were also pragmatic; that is, someone who relied on the cabañuelas needed to be a real farmer, not simply an armchair observer. As they say, *Hombre lunero no llena granero* (Men who spend too much time observing the moon don't fill up the granary), or *Labrador con mucha astronomía en eso se pasa el día* (A farm laborer spends his time in astronomy, and in the meantime the day goes by).

Though a lot of the refranes (sayings) are part of our Arab past, most have now become Christianized, that is, most now refer to the Christian calendar, though the knowledge was embedded in the Calendario de

Córdoba, the Calendario Anónoimo Andalusí, and the Tratado de los Meses of Ibn Asimor and other calendars, such as those from Yemen. For example, garlic is usually planted by San Martín, or November 11, fabas usually by San Lucas, October 18, and the winter vegetables should be transplanted by Santiago, July 25.

The name January comes from the pagan god Janus, which signifies a door separating space and time (*eones*, from the Greek *aion*). The god usually has two faces, the past and the future, sometimes a face of a man and that of a woman, representing the duality of nature. Farmers would look to the quarter moon to see whether it would rain or not. If the quarter moon was on its belly, that meant it would be dry, but if the moon was tipped toward the bottom it meant it would rain. Or in the winter, if there was a flock of crows hovering real close to the ground and crowing, it meant a snowstorm was on its way. Then people were very observant of nature because understanding nature could mean the difference between a good harvest and a lean winter.

At the beginning of the New Year there was usually some individual in the village who could "read" the cabañuelas, predicting the weather for the whole year. In general the cabañuelistas don't want rain at the beginning of the year, for it's a bad omen. They say that the cabañuela *se vacia* or *se revienta*, that is, that the cabañuela "empties" or "bursts."

In the past people were very observant of their surroundings. They would take into account the color of the sky when the sun was going down in the west. If it is light pink, pale yellow, or grayish, that usually signifies a change in the weather. But if the sky is an intense blue, it signifies heavy winds in the upper atmosphere.

Changes in the weather could be observed by paying attention to the barn animals. If the goats are eating and moving rapidly it means a storm is coming, and if the cows lie down to eat it means a rainstorm is on its way. If the rooster sings at midday or the barn animals are very tranquil, or people have pain in their joints, changes in the weather are coming, and the cats start running and jumping all of a sudden, expect wind. If the cat is washing its face or if the frogs croak louder than usual, that usually means rain. In the summer if the roses smell more intense, it means there is low pressure, but if the smoke spirals straight up in winter it signifies atmospheric stability.

When we delve into how the weather was predicted in the past, the cabañuelas, we have to reach back into the Sephardic past, since they are Jewish in origin. Jews in Toledo celebrated the Feast of the Tabernacles or Feast of the Cabañuelas in August in memory of the forty days they spent in the desert. After wandering for forty years in the desert the Jews had become very observant of which years the cows were lean and when they were fat. For no one knows exactly where the term came from, but it is believed it came from Zamuc, or Fiesta de las Suertes, the feast of luck, from the Babylonian calendar, which in Hebrew translates to the Fiesta of the Tabernacle. In some parts of the Spanish-speaking world there are the Cabañuelas de Santa Lucía, observed from the thirteenth of December to the sixth of January. The tradition of observing the cabañuelas seems to be exclusive to the Spanish-speaking world, from Spain and the Canary Islands to northern New Mexico to Mexico and Cuba. In parts of Spain, including the Canary Islands, the cabañuelas are still observed in the month of August, with the first of the month known as *llave del año*, or the "key to the year."

It was not only the Jews, Muslims, or Christians who kept track of the cabañuelas. The Muslims had their *almanaques*, "calendars," while the Christians followed the Jewish tradition of the cabañuelas, only they substituted saints days. Meanwhile, the Mayans called the cabañuelas *chac-chac* and observed them in much same manner as other groups. The ordering of months from January to December was known as *xoc-kin* and from December to January as *ualak-xoe*. It is believed that in pre-Hispanic Mexico the Aztecs adopted this knowledge from the Mayas, and it was later adopted into the Christian calendar.

Agriculturalists think that the first of January, or August, depending when the cabañuelas are read, give a glimpse as to what the weather will be for the upcoming year. If the first of January is not observed as llave del año then the cabañuelas can start on the first of the year. Otherwise the cabañuelas are started on the second of the month, with that date representing January, the third February, the fourth March, and so on, so that the thirteenth represents December. Then you start counting backward—that is, the fourteenth represents December, the fifteenth November, until you get to the twenty-fifth, which is January again. Then the twenty-sixth represents January and February, twelve hours for each day, and so on through the thirty-first, representing November and December. But if the first is not observed as the llave del año, then the thirty-first is broken into two-hour segments to represent each month.

In order to "read" the cabañuelas properly, a person has to know where the wind is coming from. During cabañuelas if the rastrojo, or stubble, on the fields is *correoso*, "flexible" or "leathery," in the morning, it means clouds; if the grapes are moist, it signifies cold.

The cabañuelas can also be said to be a ritual of creation and regeneration, for they are based on the vast amount of knowledge that the agriculturalist has about his space, especially his individual microclimate, his querencia. For example, the cabañuelas for Taos will not be the same as those for Santa Fe or Albuquerque, much less Las Cruces. Cabañuelas apply only to the microclimate where the individual agriculturalist is reading them, and it takes years and generations to acquire this knowledge, which is passed on orally.

The system is kind of complicated. Let's take, for example, the month of June. We know that the month is represented by the sixth and nineteenth of January, the afternoon and night of the twenty-seventh, and the hours of noon to 1:59 p.m. on the thirty-first. This way we will be able to know when it will rain, which months will be the hottest, when it will be cold, when it will freeze. For example, if the ninth of January reads as it being cloudy, temperate, or rainy, someone will probably say, "We are in the cabañuela of September." Then the following day might be windy and kind of cold, and someone will explain, "We are in the cabañuela of October."

Mi Querencia

Acequia Junta y Ciénaga,
a Sense of Place in a Displaced World

≈

MI ALMUNYAH DE LA JUNTA DE LOS RÍOS

NOW THE LAST LEG OF our journey brings us to La Junta, to my *almunyah de la junta de los ríos*, my 2.5 acres on this beautiful and bountiful piece of land that has been in my family since 1725. It is an almunyah in the true sense of the word, in that it is both a private experimental garden and an orchard and a recreational space. Mine is the seventh head gate or regadera on the Acequia Junta y Ciénaga and the fourth piece of land. The first irrigated parcel on the acequia is known as La Oscurana, "darkness," probably because in winter, since we are in the canyon between Velarde and Pilar, the sun sets a little after three in the afternoon. This property was owned for a long time by Father Peter Kupper, a priest who was born in Germany and immigrated to this country at the turn of the twentieth century. Father Kupper established the apple orchard in 1928, according to the late Elidio Gonzales, who helped with the planting. That property has been owned by Harvey Frauenglass for the past thirty years; several years ago he sold the orchard for $300,000 and soon after the new owner put it on the market for $650,000. Next follows the property of Clovis

Romero, followed by the property of the late Perfecto Maes, then the property of Walter Archuleta. The Romero, Maes, and Archuleta properties, as well as my own, at one time all belonged to my great-grandmother Ramona Archuleta and her first husband, Francisco Martín, a descendent of the original Francisco.

My piece of land has some joya-jolla and also vega and ciénaga. It does not have an altito, but the original Archuleta property had a portion of altito; today that altito is on the Romero and Kupper properties. For the acequia water to get to my land, it has to travel about eight hundred feet. The water travels through a cequiecita called a lindero, and when it gets to the end of what today is Barbara Morgan's property (at one time it belonged to my father), it drops onto a cabecera, or a horizontal cequiecita, which also serves as a desagüe, and it goes all around my property, crosses State Road 68—the main thoroughfare between Santa Fe and Taos—and eventually empties back into the Río Embudo, about two hundred yards from where that river empties onto the Río Grande. On my cabecera I had six compuertas (gates) installed in order to irrigate all my land. I still need about two more, one of them on the east side of the property. Once I realized that the land, like us, also has a memory and one probably that is longer than ours, I found there were several bancos, or step terraces, on my property. My grandfather more than likely carved out these terraces, or they might have been there since before his time. In 2012 I installed underground piping all along the lindero and cabecera, about 1,500 feet, since it was a long haul and gophers kept spreading the water in every direction, causing problems for my neighbors. Now I have eight ten-inch alfalfa valves and don't waste water as before, though now I have to reconfigure part of the landscape again.

To do justice to the original land and understand its layout, it's better to see the land as a whole. Let's start by defining an almunyah. In his *Tratado de agricultura*, Ibn Luyun, the last of the Spanish Arab agricultural writers, could well have been describing a traditional *jardin nuevomexicano* when he described an almunyah:

With regard to a house set amid gardens, an elevated site is recommended, for reasons of both vigilance and layout. And let them have a southern aspect with the entrance at one side, and on the upper level the pool and well; better, instead of a well have an acequia (waterway) where the water runs underneath the shade. And the house should have two doors, so that it will be better protected and easier for the repose of its occupant.

Then next to the pool plant shrubs whose leaves do not fall and therefore rejoice in the sight. Somewhat further off, arrange flowers of different kinds, and, further off still, evergreen trees. Around the perimeter plant vines, and in the walkways that crisscross the enclosure, a sufficiency of climbing vines.

The garden should be surrounded by one of these walkways whose object is to separate it from the rest of the property. Among the fruit trees, besides the vineyards, there should be lotus trees and other, similar trees, because their wood is beneficial.

At a certain distance from the vineyards, what is left of the farmland should be for planting and it should be favorable to whatever is planted.

In the background plant fig trees and other, similar trees. All the big trees should be planted to the north, whose purpose is to protect the rest of the property from the north wind. In the center of the property there should be a summerhouse pavilion, dowered by places to sit, with vistas to all sides, but of such form that no one approaching can overhear a conversation within and whereunto none can approach undetected. Climbing roses, as well as big myrtles, and all the plants common to a fruit and flower garden should surround the pavilion. It should be longer than it is wide, in order that the beholder's gaze may expand in its contemplation.

But a northern New Mexico landscape also resembled a Greek garden. It is said that "el huerto de los Griegos era un jardin de frutales y hortalizas donde tomar el fresco en el atardecer bajo una parra" (A Greek orchard-garden was a garden of fruits and vegetables where one could enjoy the cool evening under the grapevine). An Islamic garden has also been described thus, again very northern New Mexican: "El paraiso del Islam es un jardin donde el agua que corre y baila, refresca y el olor de las flores se esparce, impulsada por el sol y recogida por la sombra" (The Islamic paradise is a garden where the water that runs and dances refreshes and the scent of the flowers is spread, prompted by the sun and gathered by the shade). Thus the New Mexico landscape is influenced by these two great cultures, for almost every historic landscape is a combination of the two. On a hot summer day this type of landscape is transformed into *el frescor del paraiso* (the freshness of paradise). If in the winter and early spring the Indo-hispano looks for a resolana, a spot facing south-southeast, to enjoy the warmth of the day against an adobe wall, during the summer there is nothing better than the wonderful shade provided by a big tree. That is paradise. New Mexican folklore— which to me is traditional ecological knowledge—reminds us that in summer "es mejor una larga sombra que una corta resolana" (a long shade is better than a small amount of intense heat [a resolana]).

The Acequia Junta y Ciénaga is the last acequia on the south side of the Río Embudo and at present it irrigates approximately 80 acres, though not all the land is presently under cultivation. According to documents at the Office of the State Engineer in Santa Fe, at one time there were 200 acres under irrigation. But where are those other 120 acres? I haven't been able to figure out. The Tewa Basin study of 1935 lists only 80 acres, the same as today. At present there are thirty-seven parciantes, though some absentee landowners don't really care for the land. This creates problems with overgrown weeds and invasive species that are taking over the land and from there spreading to all the properties.

LA QUERENCIA: SENSE OF PLACE

The crucial and perhaps only all-encompassing task is to understand place, the immaculate specific place where we live. The kinds of soils and rocks under our feet; the source of the waters we drink; the meaning of the different kinds of winds; the common insects, birds, mammals, plants, trees; the particular cycles of the seasons; the times to plant and harvest and forage—these are the things that are necessary to know.

—KIRKPATRICK SALE

We started with a global perspective of water and community in arid lands, similar to the landscape of the southwestern United States and specifically of the Río Arriba bioregion of northern New Mexico and southern Colorado. To be even more specific, we focused on the area from La Bajada, about twenty miles south of the capital city of Santa Fe, to about thirty miles north of the New Mexico–Colorado state line in the San Luis Valley. To better understand the concept of water from an Indo-hispano point of view, we must go back several millennia in the history of the Middle East, Africa, the Mediterranean, Spain, Mexico, and of course New Mexico. To understand place, la querencia, the earth and pebbles, which countless generations have stepped on since people started working the land, our land today, we need a global perspective. (For we do not inherit the land from our parents; we have borrowed from our children, or so say our Native American ancestors.) And that is how this work began.

For Indo-hispanos and other Latinos throughout the Americas, such agricultural traditions mean returning to our roots, our querencia, from which our land ethic evolves. On a Latin American blog querencia was defined as "una especie de medida agraria, algo como una hectaria, un acre . . . unidad de medida" (an agrarian measurement, like an hectare, an acre . . . a form of measurement).

First, we need to understand the earth, and for that we have to re-move ourselves, that is, place ourselves above to get a bird's-eye view of our planet. At the same time we have to go back at least ten thousand years, spiral backward and upward, to understand our relation to place in space and time. If we don't get the broader picture, we will never be able to comprehend where we stand today in terms of place.

Though undoubtedly the pobladores who came with Oñate learned much about the terrain of the new land from the Pueblo and Meso-american people, their traditions regarding agricultural and irrigation techniques were Mediterranean and Middle Eastern in origin more than Christian influenced. The Moors came to Spain from lands where irrigation was practiced (Egypt, Syria, Mesopotamia, Yemen, and Jordan). Nostalgic for their homeland, they wanted to replicate their "habitats," or the agroecosystems of the Middle East, in the Iberian Peninsula.

They saw the possibility of re-creating these ecosystems of irrigation in their adopted homeland, modeled on the infrastructure that was established by the Romans but in a dilapidated state by the time the Moors arrived in AD 711. Also, the Roman system of aqueducts was more suited for urban settings, while the Moorish systems were developed more for rural areas.

Like the settlers who came to New Mexico under the Spanish Crown, the Moors settled in regions that must have reminded them of home: the Syrians established themselves in Valencia, Sevilla, Niebla, and Granada; the Palestinians in Algeciras; those from Jordan in Málaga; the Egyptians in Murcia; those from Yemen in Zaragosa, Alicante, Elche, and Nevela; and those from Mecca in Córdova. In general they settled in the fluvial plains of the most important rivers in Spain. Similarly, Oñate settled at the confluence of the Chama and the Río Grande.

In 1987, when my daughter Única Paloma Lucía was born, I embarked on a project that, little did I know, would become a lifelong challenge and adventure. It started as a very simple project. I wanted to plant some red raspberries so that my daughter could eat healthy. First I went to the local county extension service to ask their advice about the kind of raspberries that would grow best in La Junta. But I wasn't ready for their answer. "If I were you I wouldn't waste my money," I was told. "Raspberries don't do well in the Española Valley." That inspired me to try to grow them, for I never like to be told "it can't be done." Stubborn as I am (I guess that's part of being a Virgo), I sent for ten heritage raspberry plants from a mail-order catalog. I thought to myself, "If they don't grow, I won't lose much." Since I was a young kid my mom had always let me order some of the weird, or should I say different, vegetables when she would place her order for seeds in late winter. Though she saved her seeds for chile, white corn for chicos and posole, and blue corn for atole, she would always order sweet corn from the catalogs that started arriving in the mail after Christmas. For several years we ordered a Fourth of July sweet corn, and we did eat sweet corn by early July. She also ordered carrots, radishes, cabbage, and lettuce. Watermelons and cantaloupes came from seeds she saved, and also tomatoes.

Though I was expecting failure with my raspberries, lo and behold, they produced a few berries that first year, toward the end of September and into October. I would take my daughter to see the berries and put them on her lips, and though she couldn't talk, she would get all excited and move her tiny arms in approval. That spring, when she was about a year old, I would hold her in my arms and to my surprise she would make a lot of noise and point her little fingers in the direction of the plants, as if she knew where the berries were. Well, this little experiment led me to think that I needed to redesign my 2.5 acres as we have. And to my surprise I noticed what I had never seen before: there seemed to be traces of terraces. That's when it hit me that the land, like a book, can be read if one pays attention. Where I had only seen what appeared to be slopes with arroyos developing on several places on the land, all of a sudden the land revealed itself and I started to see what appeared to be terraces.

Most of my life I have enjoyed hanging out with people older than me, probably because I grew up as an only child, though I was the youngest of twelve siblings. After I was born my biological mother was diagnosed with cancer. Several of my aunts volunteered to look after me while my mother recuperated. My tía Roberta already had a large brood of her own, so I ended up with the younger brother of my biological dad, Carlos, and his wife, Lucía, who didn't have kids of their own. What a blessing. I could never have wished for better parents. So though I was the youngest in my biological family, I grew up as an only child. My oldest brother, Adolfo (named after my biological father), was born in 1928. The second oldest, Wilbert, drowned at the age of twenty-two in the Río Grande—in La Nasa. Then came Bialquin, followed by my oldest sister, Alice, then Tomasita, Elise, Nelson, Arnold, another Estevan who died as an infant, my youngest sister, Celia (named for my biological mom), and then I was the last. I was named to replace my brother, who was named for my uncle Juan Estevan, who died in 1944.

For the first seven years of my life we lived in Cañoncito, the uppermost hamlet within the Embudo land grant, where we had about five acres of land, the lower half irrigated by the Acequia del Medio and the upper half by the Acequia de la Sancochada. In Cañoncito we lived on the land that had belonged to Isidro Martinez and Alvina Maes, my great-great-grandparents, the parents of María de la Luz Martinez, who had married Jose Ignacio Arellano and had given birth to my grandfather, José Agustín Arellano, in 1868. Isidro and his brood owned most of the land today watered by the Acequias Leonardo Martinez, Duranes, Sancochada, and Medio.

In Cañoncito we had a lot of fruit trees because my father (Carlos, as I always called him) loved fruit trees. At one time he had about fifty-five sweet cherry trees and sixty pear trees of different varieties, as well as apricots, peaches, and grapes. Among them were what he called *fruta americana*, plus *manzanita, cerezos,* and *albaricoques mexicanos.* The so-called Mexican fruits were the very old varieties that today are known as heritage fruits. He also had the ones he had ordered from his "primo Benjamin Atencio," who used to take orders for American-type varieties from Stark Brothers Nursery by going house to house, taking orders.

At that time I knew nothing about terraces, but I first heard from my father the words that developed into the vocabulary I have spent the past twenty-plus years trying to understand and define. From him I first heard about the "altito," the "joya-jolla," "vega," "ciénaga," the "tira," and so on. Without knowing it, he influenced the way I still look at the acequia landscape today, as I grew up listening to him and his contemporaries define the land in Spanish, the New Mexico Spanish that was their native tongue and also mine. I didn't speak a word of English until I went to school, and then only in class. During recess and in the mornings and after school our primary language was Spanish, and for me it still is at home.

In 1954, due to water shortages in Cañoncito, we moved to where we still live. This place is known as Embudo because of the post office, but the locals refer to Junta and Ciénaga and, across the river, Rincón (the inside corner; in Spanish the outside corner is known as *esquina).* The reason we moved was that my dad was tired of getting up in the middle of the night to irrigate by the light of a *farol* (lantern) because in most years, during July and August, people in the four upper acequias had to share the water (repartimiento). The four upper acequias, two on the north side (Leonardo Martinez and Duranes) and two on the south side

(Sancochada and El Medio), had the water for three days, then the four lower acequias, La Plaza and El Llano on the south side (incidentally the two oldest within the land grant on the south side of the Río Embudo) and La Apodaca and El Bosque on the north side, had the water four days. I think this is where my father learned to water with an *hilito de agua*, a "trickle of water"; necessity showed him how to conserve water.

Then, in 1964, coming home from work in Los Alamos, my father was injured in an automobile accident. He was hit by two ladies from Los Alamos who were driving drunk, which left him in a coma for over a month. (I was in school at McCurdy at the time of his accident.) He never recuperated and was unable to do any work like before. He died from a massive heart attack in 1978 at the age of seventy-one. They finally sold the land in Cañoncito in 1966, leaving us with only the land in La Junta. This land was part of the original land of my great-grandmother Ramona Archuleta, the mother of Tomás Archuleta, father of Celia, my biological mother. Ramona and her two brothers seem to have owned most of the land that we know today as La Angostura, Bolsa, Rincón, Junta, Ciénaga, and La Mesita. Ramona married a man forty years older than her (she was twenty and he was sixty; he died at age seventy). Ramona also had two daughters, who ended up marrying two brothers: Alcaria married Matias Romero and Donaciano married the younger sister Benigna. Recently I came upon a document showing that Francisco Martín, a descendent of the original Francisco who was granted the Embudo grant in 1725, gave the land from "los Barrancos blanco" to "la Angostura" to his daughter, Alcaria. An uncle used to say that their last name should have been not Archuleta but Borrego, and he was right. The genealogist Henrietta Martinez Christmas found a document indicating that my grandfather's father, Ramona's second husband, was named Juan Antonio Borrego. Borrego probably died when my grandfather was an infant, and my grandfather, Tomás Aquino, later adopted Archuleta as his last name.

Today La Mesita has several owners, one of whom spent over $100,000 to build a road up to the mesa where they hope to build their house. When it was first announced that they had a acquired the 640 acres (one section), they wanted to build twenty-four houses on it. And they probably will, once the opposition settles down. That section at one time belonged to my grandfather Tomás. Today the other part of the mesa belongs to the Bureau of Land Management and to Rudy Jaramillo; his father, Juan, used to run sheep in the winter and spring. On the south side of the mesa ran the original Camino Real de Tierra Adentro between Ohkay Owingeh and Taos to the north and Picurís to the east. It's also here that the infamous Battle of Embudo took place on January 29, 1847, during what's known as the Taos Rebellion, part of the Mexican American War in which Mexico lost about half of its territory to the United States upon the signing of the Treaty of Guadalupe Hidalgo on February 2, 1848.

To put everything in context, my great-grandmother Ramona was born in 1822, during the Mexican period, which made her a "Mexicana," while my grandfather Tomás was born in 1856. His first wife was María Marcelina Salazar, whom he married in 1876 and who died in the late 1890s, after giving birth to twins who apparently didn't survive. He then married Eva Duran, who gave birth to my biological mother Celia in 1906. From his first wife, Tomás had a daughter named Deluvina, who was the mother of Lucía, who raised me. Deluvina married Rafael Borrego (Rafael and Tomás had the same grandparents—Luis Pino Borrego and Antonia Severina de la Cerda) from El Guique, and Lucía was born in 1913. This made Lucía the niece of Celia, my mother. Celia married Adolfo in 1926 and Lucía married Carlos in 1930 in Chamita. Adolfo and Carlos were brothers, and they had another brother, Delfino, and three sisters: Eloisa, who died in childbirth at a very young age; Delfina, who married Fidel, Eloisa's widowed husband; and Merced, who married Rosinaldo Romero from Ojo Sarco, on the eastern part of the Sebastián Martín land grant (later part of the Trampas grant). The communities of Río Lucio, Vadito, Peñasco, and parts of Chamisal, though Indo-hispano communities, are all located within the historic Picurís land grant.

This is a very brief outline of who I am. On both sides of my family I am related to Francisco Martín, who is a descendent of the Martín Serranos who first came to what is now New Mexico in 1598. The Martín

Serranos are first found in Mexico City in 1525 and in Zacatecas by 1575. The Arellanos didn't come to New Mexico until after the Pueblo Revolt of 1680; Juan Cristobal Arellano came with don Diego de Vargas during the so-called reconquest of 1695 from Aguas Calientes. The Arellanos originally came from the town of Arellano in Navarro in the Basque Country, about eighty miles from Pamplona, Spain, where the bulls are run every July for the Fiesta of San Fermín.

This is the lineage of my almunyah de la junta de los ríos. The tangled history of my family shows that life in northern in New Mexico was very complex. We are a mixture of blood from the Iberian Peninsula, Basques and Sephardics and more than likely Moors who mixed here with Meso-americans, then Pueblos, Apaches, and Navajos, in the case of my kids. All these bloods informed us about how to look at the land and water so that what we have today is something new, what José Vasconcelos called La Raza Cosmica. A new landscape based on "el juicio de la tierra y la sabiduría del agua."

Epilogue

A la acequia o a la escuela: Acequia Literacy

≈

Our journey is now complete, and we now know the truism of the dicho *Sin agua no hay vida* (Without water there is no life). If we want to continue the traditions of our ancestors, from the Indus Valley to Embudo, water and community have to remain one. Water cannot become a commodity, for today more than ever we have to remember the Muslim Law of Thirst, that water belongs to every living thing, be it a plant, an animal, or a human being. In our journey we encountered a lot of traditional knowledge and saw how, when that knowledge was lost, the civilizations that depended on it crumbled. Look no further than the sacred valley of the Incas, the chinampas of Xochimilco, Ses Feixes in Ibiza, and the hundreds of irrigation canals that have reverted back to desert. When we forget that sacred knowledge that bonds water to community, humanity disappears. And the acequias are in a very precarious state at this time. Either we recoup this *oro del pueblo* or in a few years they will be but a memory.

Either no one is preoccupied about the acequias and other community irrigation systems or there is a need for a reinterpretation of the landscape and water. Náyade Aguirre, in a column for *El Observatodo* in Chile, writes,

The development and growth of our city has permitted the ancient agricultural lands to be turned into towns, condominiums, and villas. But there have remained on its periphery trenches of the ancient

acequias and canals that today are only receptacles for trash and rubbish. The trenches are full of junk that deface the city and serve as trash
collectors for those insensitive neighbors who prefer to travel for miles
and deposit their trash and rubbish (when logically they should simply
take out their trash to the door of their house on the day of trash collection). These trenches are not only a sanitary danger or an ugly site
but rather represent a great danger to the security of the community,
since they are used as a refuge for delinquents.

She continues,

> To whom do these pieces of land belong? Do they belong to those who
> own the canals even though they don't deliver water for irrigation? It
> seems important to have ordinances pertaining to construction, so
> that in each new development . . . whether as a new green zone or
> another alternative, in accordance with the neighbors, [the canals
> are] maintained if possible for irrigation purposes. The construction
> companies buy agricultural lands next to these lands [former ace-
> quias or canals], and that's why when building they are not incorpo-
> rated, but it is the municipality (I think) that should demand a
> solution for these ancient irrigation structures that are not in use, so
> that they don't become a community problem. If a canal or acequia is
> not in use, the owner should be required to . . . maintain it.

Traditional agriculture is no longer a producer of food, as in the past,
but rather it provides other services for society, such as the maintenance
of the agrarian landscape, the preservation of the environment, and the
provision of green belts.   •
    The new role of traditional agriculture in today's world is the conser-
vation of the environment and its biodiversity and the occupying and
rearranging of the landscape; today this is known as multifunctional
traditional agriculture. The production of food should again become a
priority, however, if not for the market then for local consumption. The
population has to be *concientizada*, as Paulo Freire would say, about the
positive aspects of a traditional agricultural landscape, whether it's an
acequia landscape or the chinampas of Xochimilco.

Trash in the acequia. Photograph by Donatella Davanzo.

INTERPRETATION OF THE LANDSCAPE
AND THE WATER

Traveling through rural communities, whether in New Mexico, Spain, or Mexico, makes us reflect on the sentiment of querencia toward the landscape. Today traditional rural agricultural landscape environments have been damaged in the process of accelerated transformation, causing them to lose their scale and their appearance as a rural scene.

There is a possibility of preserving certain landscapes, particularly those of great value and those in the most alarming processes of decay. To protect others from extinction there has to be radical change. This preoccupation with the landscape must be translated into an elaborate series of public policy alternatives by different governmental administrations—local, state, regional, or national—in an attempt to visualize the importance of this landscape for the development and well-being of society.

There has to be a dignified future for the agrarian landscape. We must create a "singular agrarian landscape" as a protected landscape, a multifunctional, biodiverse landscape, by providing social and institutional support in order to enter a new era.

First, we have to understand the importance of the rural landscape within acequia culture. Each landscape and the elements that make it up are unique and irreplaceable, since they identify the territory and its population. In that sense the patrimonial value of the landscape is unquestioned and it is seen as an important resource within the territory.

Second, we must understand the landscape as a result of a compromise between culture and nature. Landscape cannot exist without humans. Remember that acequias are artificial, man-made environments that mimic nature. The landscape is nature trampled, where there coexist ways of exploiting the land and different ways of managing property, roads and trails, sprawl and clustered housing, rivers, ravines, mountains, forests, and so on. Any human actions transform the landscape.

Third, we have to assess the role of politics in the revaluation and renovation of the rural patrimony, with special attention to the singular agricultural landscape. In this context agriculturalists are important actors in terms of preserving the quality of the landscape.

If there is a change in any of our cultural premises, the landscape changes, and we then realize how fragile the rural landscape is in the face of the transformations of our contemporary civilization. A long series of local planning and model ordinances have attempted to save this rural patrimony, but their results have not been very fruitful in northern New Mexico.

What we need to do is inventory the different types of agricultural landscapes and bring to light the typical rural architecture, such as the acequias and desagües. We need to find ways of conserving the landscape, including the flora and fauna as well as the role the agricultural landscape has played in the evolution of the surrounding area. There also has to be an analysis of the possible future uses for such a landscape, taking into account new technologies. Before we abandon the past (flood irrigation) for the contemporary (drip irrigation), we need a thorough analysis of the pros and cons of each system for the whole cultural landscape. The future may be one where the old and new learn to coexist, such as the hoe with the plow.

The public, including agriculturalists, politicians, developers, and law enforcement officials, has to be educated about the importance of such a landscape. And the results of the project have to be disseminated in order to start a dialogue.

The ultimate goal should be the conservation of the agricultural landscape, including saving and propagating at-risk plants and trees. This can be accomplished by generating materials for the general public to interpret and promote the value of the disappearing agricultural landscapes; sponsoring local art and photography contests; making information available in local schools so youth can learn the value of the agricultural landscape; and holding workshops and community forums to discuss ideas and gain support for the preservation of the agricultural and acequia landscape.

The landscapes visited above have to be put back into circulation, revived, and returned not only as agricultural producers but also recreation spaces. We need more almunyahs everywhere, especially with climate change approaching, in order to experiment with new ways of adapting crops to the changes in weather patterns. We need a new landscape where rural and urban meet, based on traditional knowledge yet ready to test and accept new technologies that don't contribute to its destruction. We are entering a new frontier if we are open to learning from the past and preparing for the future and its approaching changes. Just as there was an agricultural revolution in AD 711 when the Moors crossed the Strait of Gibraltar to the Iberian Peninsula, or when Columbus encountered a thriving civilization upon crossing the Atlantic, today we are at a new threshold. Either we continue with this experiment with moving humans forward, or we perish. From all indications we will change and the landscape will be our ally if we follow the wisdom of the land and the knowledge of the water.

Note on Sources

≈

In writing this book I have drawn on a variety of my own previous research and scholarship.

In chapter 3, in my discussion of the Sebastián Martín Land Grant, much of the material is taken from "*La Cuenca y la Querencia*: The Watershed and the Sense of Place in the *Merced* and Acequia Landscape," in *Thinking Like a Watershed: Voices From the West*, edited by Jack Loeffler and Celestia Loeffler (Albuquerque: University of New Mexico Press, 2012), 151–91; and "Land Use and Water History," which I wrote for the 2007 *Upper Río Grande, New Mexico: Rinconada, Embudo, Velarde and Alcalde Watershed Management Plan*, prepared by Environmental Health Consultants (Embudo, NM).

In chapter 4, I used some of the following materials that I published in *Green Fire Times*: "There is no food security, without acequia security," *Green Fire Times* 3, no. 3 (March 2011); "Understanding 'Cooperación,'" *Green Fire Times* 3, no. 1 (Jan. 2011); "Cabañuelas: Jewish and Mayan Meteorology Traditions Combined," *Green Fire Times* 2, no. 12 (Dec. 2010); "All Española Valley Chiles Are Related," *Green Fire Times* 2, no. 11 (Nov. 2010); "La Abundancia del Jardín Rizo: The Bounty of the Garden," *Green Fire Times* 2, no. 9 (Sept. 2010); "Developers and Landowners Should Take a Page From the Past," *Green Fire Times* 3, no. 9 (Sept. 2011); "Designing an Edible Landscape," *Green Fire Times* 3, no. 12 (Dec. 2011); "La Milpa: A Sustainable Model," *Green Fire Times* 4, no. 1 (Jan. 2012); "Convide: A Sustainable Philosophy," *Green Fire Times* 2, no. 10 (Oct. 2010); and "Una Vida Buena y Sana y Alegre—A Sound, Healthy (and Cheerful) Life," *Green Fire Times* 4, no. 2 (Feb. 2012).

Additionally, I drew from "Designing Acequias That Deliver," *Taos Green Guide* (Jan. 2008): 43–45; "Acequia Democracy," *Santa Fe Resource Guide* (2007): 43–45; and "Acequias—The Way of the Water," newmexicohistory.org.

In chapter 5, I used some of the material I had written for "Taos: Where Cultures Met Four Hundred Years Ago," *GIA Reader* 18, no. 3 (Fall 2007): 49–56; and "Querencia: The Soul of the Paisano," *The Quivira Coalition Journal* no. 31 (September 2007): 9–14.

Index

≋

Page numbers in italic text indicate illustrations.